Multilingualism Online

By the co-author of *Language Online*, this book builds on the earlier work while focusing on multilingualism in the digital world. Drawing on a range of digital media – from email to chatrooms and social media such as Facebook, Instagram, and YouTube – Lee demonstrates how online multilingualism is closely linked to people's offline literacy practices and identities and examines the ways in which people draw on multilingual resources in their internet participation. Bringing together central concepts in sociolinguistics and internet linguistics, the eight chapters cover key issues such as:

- language choice
- code-switching
- identities
- language ideologies
- minority languages
- online translation.

Examples in the book are drawn from all the major languages, as well as many lesser-written ones such as Chinese dialects, Egyptian Arabic, Irish, and Welsh. A chapter on methodology provides practical information for students and researchers interested in exploring online multilingualism from a mixed-methods and practice-based approach.

Multilingualism Online is key reading for all students and researchers in the area of multilingualism and new media, as well as those who want to know more about languages in the digital world. Additional resources for Language and Communication Studies are available on the Routledge Language and Communication Portal: http://www.routledgetextbooks.com/textbooks/languageandcommunication.

Carmen Lee is Associate Professor, Department of English, at the Chinese University of Hong Kong. She is the co-author of *Language Online: Investigating Digital Texts and Practices* (Routledge, 2013).

Multilingualism Online

Carmen Lee

Routledge
Taylor & Francis Group
LONDON AND NEW YORK

First published 2017
by Routledge
2 Park Square, Milton Park, Abingdon, Oxon OX14 4RN

and by Routledge
711 Third Avenue, New York, NY 10017

Routledge is an imprint of the Taylor & Francis Group, an informa business

British Library Cataloguing in Publication Data
A catalogue record for this book is available from the British Library

Library of Congress Cataloging-in-Publication Data
Names: Lee, Carmen, author.
Title: Multilingualism online / by Carmen Lee.
Description: Milton Park, Abingdon, Oxon ; New York, NY : Routledge, [2016] | Includes bibliographical references and index.
Identifiers: LCCN 2016020423 | ISBN 9781138900486 (hardback) | ISBN 9781138900493 (pbk.) | ISBN 9781315707211 (ebook)
Subjects: LCSH: Language and the Internet. | Multilingualism—Technological innovations. | Language and languages—Technological innovations. | Communication and technology. | Multilingual Web sites.
Classification: LCC P120.I6 L37 2016 | DDC 404/.20285—dc23
LC record available at https://lccn.loc.gov/2016020423

ISBN: 978-1-138-90048-6 (hbk)
ISBN: 978-1-138-90049-3 (pbk)
ISBN: 978-1-315-70721-1 (ebk)

Typeset in Times New Roman
by Apex CoVantage, LLC

Contents

Figures

Tables

Acknowledgments

This book would not have been possible without the many people who offered their professional, editorial, and moral support throughout the whole project.

First and foremost, I would like to thank David Barton for encouraging me to write this book and for his helpful guidance and comments on earlier versions of the manuscript. I also thank the reviewers of the book proposal and those who shaped my initial thinking of the book, especially Caroline Tagg and Camilla Vasquez.

I am grateful to have the book introduced by Mark Sebba (Lancaster University) and concluded by Susan Herring (Indiana University), both of whom have hugely inspired my work over the years. Mark and Susan also offered their insightful comments in addition to writing their pieces, for which I am very thankful.

I must acknowledge the significance of *The Multilingual Internet* in 2007 (Oxford) co-edited by Susan Herring and the late Brenda Danet. The volume, being the first of its kind, opened a window into the world of internet multilingualism. Being able to write my own book on the topic a decade after the publication of *The Multilingual Internet* means a great deal to me!

I am indebted to the Chinese University of Hong Kong, especially colleagues in the Department of English, for granting me a sabbatical leave from September to December in 2015. With this time off, I was able to plan this book and to fully concentrate on my research and writing. I also visited my alma mater, Lancaster University, in November 2015. In particular, I thank Julia Gillen and members of the Literacy Research Centre at Lancaster for inviting me to deliver a research seminar. I also thank them for their constructive suggestions and comments on my talk.

Thanks must also go to my current and former research students and assistants who contributed to the various research studies covered in this book. For this book, I especially wish to thank Smile Xiao for her meticulous editorial assistance.

Of course, this book would not have become a reality without a supportive publisher. I am grateful to Louisa Semlyen of Routledge for wanting to publish this book in the first place and to Laura Sandford for her professionalism and patience.

I would like to dedicate this book to my mother, who has always believed in what I do, and to my family and friends who stand by me and encourage me in everything. Do ze saai!

Every effort has been made to contact copyright holders. Please advise the publisher of any errors or omissions, and these will be corrected in subsequent editions.

Foreword

In the early days of the internet, it was often observed that electronic communication (which in those days really meant email, or e-mail as we used to write it) had many resemblances to speech which made it contrast with traditional styles of written communication. In particular, freedoms of expression, such as informal styles of address and abbreviation and that which were characteristic of spoken, face-to-face conversation, seemed to be allowed in computer-mediated writing in a way not tolerated in old-fashioned writing on paper.

Despite that, it seemed, there was one respect in which language on the internet might not resemble spoken language. While there were thousands of spoken languages, it looked as though the internet would be home to only a few of these, with English being overwhelmingly dominant. Furthermore, while bilinguals and multilinguals could mix languages and engage in other kinds of multilingual practice in speech, it was not clear that the internet was going to make this easy, or even possible. The internet seemed, for a time, to bear the promise of a future where communication would be very easy, fast, cheap, and relaxed, but at the same time it appeared to threaten a monolingual apocalypse where the languages of a few technologically advanced and economically important countries would prevail, to the exclusion of others.

It didn't quite happen like that. As Carmen Lee's book shows, a decade or two later, "multilingualism" is thriving on the internet. Improvements in language-related technology (for example, the provision of fonts in non-Western scripts and the availability of moderately good machine translation), a new wave of internet affordances (Web 2.0 with a huge range of synchronous and asynchronous interactions on offer), and the creativity of millions of users have turned the internet into a very multilingual place. This multilingualism is, however, not multilingualism exactly as we used to understand it. Slowly, and only recently, through the work of the late Jens Normann Jørgensen and many others, it has become clear that most if not all of us humans can be a little bit multilingual, engaging in the practices which are now known as *polylanguaging* and *translanguaging*, even when we don't feel we know any languages apart from one first language. An open space like the internet is an ideal place for such practices to go on, although they almost certainly did not start there. As offline research has expanded our understanding of these practices, the internet provides a rich site for studying the development of their online forms.

As Carmen demonstrates in this book, multilingualism on the internet is much more than just the use of two or more languages. It encompasses both the kind of multilingual practices familiar from the predigital age and new practices in which even monolinguals can engage with people whose languages they barely know or don't know at all (for example, through the use of online machine translation). Furthermore, the internet provides informal spaces where such tentative bilinguals, as well as more fluent ones, can reflect on and be supported in their language learning and language use. Carmen also shows how multilingualism, whether as a fluent speaker/writer of languages, as a language learner, or as a monolingual participant in other people's multilingual practices, can be a powerful component of online identities.

Carmen has been comprehensive in her approach. *Multilingualism online* could be taken to mean a variety of things, such as the use of different languages on the web in general, websites which contain pages in more than one language, code-switching, translanguaging, and translation. This book covers not only all of these, but also how multilingualism is *talked about* on the internet ("thematization") and how it is researched. The discussion of methodology is particularly useful, because despite the internet having been with us for some decades, the most fruitful ways to research it are still matters of discussion – and all the more so when multilingualism is the focus.

Carmen's book is a scholarly and timely contribution to the study of multilingualism in the world online. Its overview of research in all the areas mentioned is thorough. But because of the way she focuses on *practices*, the online world and the offline world are never that far apart. As Carmen says:

> What I do online is tied closely to my offline lived experiences. . . . This also means that research into language on the internet must take into account not only texts on the screen but also what people do with these texts in other areas of their lives.

That connectedness, in my mind, is the great strength of this book, and it is one of things that makes it a rewarding read. Even those of us who live our lives rather monolingually are linked, through the internet, to a polyglot world. Next time I go to Facebook, I may be confronted with a post in a language I know slightly or not at all. I can then choose to ignore it, to use the automated translate function, or to go to a website like Google Translate for a rough translation. I could also ask someone in the next office or turn to a dictionary on my shelf or in the library. If I decide to respond, I could use a similar strategy or (still fairly safely) add a comment in English, knowing that it is likely to be understood and, even if it isn't, it will be translated by some means, online or off. Thus, online practices link to offline practices, though not always in predictable ways. This book will enrich our understanding of the multilingual online world, but equally, it will add to how we understand the multilingual world offline.

Mark Sebba

1

BACKGROUND AND APPROACH TO MULTILINGUALISM ONLINE

Overview

- Multilingualism online: An auto-technobiography
- Why multilingualism online?
- Beyond multilingualism
- A practice-based approach to multilingualism online
- Overview of chapters

MULTILINGUALISM ONLINE: AN AUTO-TECHNOBIOGRAPHY

My very first experience of personal computing dates back to the early 1990s at home in Hong Kong, when I was still a high school student. I remember the first thing I did on the computer was play a card game called Solitaire. At that time, I had to share a desktop computer at home with my two younger brothers. After a few years, we also had an internet connection, but I did not take much notice of it. I remember it was my brothers who were always typing something on a black screen, but I had no idea what exactly they were doing. I was, however, sure that they were typing something in English, which was quite strange to me as they rarely used English (except for school work). They later told me that they had been communicating and exchanging files with people from other parts of the world through a bulletin board system (BBS). I was very impressed with what they could do, but I had no intention of learning more about it because I only used the computer occasionally to format my assignments. (Typing up homework was still optional then, but I thought a word-processed piece would make a better impression!)

One day a high school friend who had moved to Australia asked me if I had an "email address" as she would like to write me an "email" – a completely new idea to me. With some help from my brothers, I finally managed to write and send my very first email. It took me a whole afternoon to compose it. I typed it in English because Chinese was almost impossible for me (processing nonalphanumeric characters was not easy then). That very long email, as I recall, closely resembled a formal business letter that I would have written for an English composition class!

Sometime in my last year in high school, I first came across ICQ, an instant messaging program. I was extremely excited about being able to communicate with people in real time by simply typing on the computer, even though we could not hear or see one another. My very first ICQ message sent was a simple "Hi" to my cousin. Because dial-up internet service was quite costly, I only went online to chat with friends for a very short time each day. I still preferred to type most of my messages in English only (rather standard or formal English, and sometimes with a few emoticons here and there). I had learned some Chinese typing, but I was never good at memorizing the codes. When it came to surfing the web, the only things I did were read the news and look up materials for my assignments. While most websites I came across had only English content, I began to notice that more and more webpages were available in multiple languages. Tools such as free online dictionaries and translators also emerged, and I still remember my teachers always warning us about how unreliable some of these tools were.

The internet gradually gained its popularity in Hong Kong in the late 1990s, when I was an undergraduate student. At university, all students were given free dial-up access to the university internet servers, with limited monthly connection time. Surfing the web, emailing, and chatting on instant messenger (IM) at the same time gradually became a habit. This was also the time when my parents gave me my first mobile phone, though I used it for calls only, as texting was quite costly then. And when I did text, English was still my preferred language.

Later, my IM activity switched from ICQ to MSN messenger (later called Windows Live Messenger). I noticed that on MSN, I no longer wrote my messages in English only; with improved technologies, I felt quite at ease playing with the different languages and scripts available to me. Cantonese is the major everyday spoken language I use with my family and friends. I learnt English in kindergarten, in primary school, and through my high school years. In high school, English was the medium of instruction for non-Chinese subjects. At university, I studied English and linguistics. These subjects also provided me with many opportunities to read and write in English. Outside the university, I communicated with others mostly in Cantonese. I had learnt some Putonghua (Mandarin Chinese) in primary school, and I took French as one of my undergraduate courses. Because Cantonese does not have a standardized writing system, in Hong Kong, standard written Chinese, represented by traditional characters, is adopted as the common written language in Chinese.

My linguistic background significantly shapes my deployment of languages on the internet. I am aware that my choice of language online changes from time to time. I used to chat in a relatively standard form of English, although it was not my main language elsewhere. I also started using more Chinese characters in IM when I learnt Chinese inputting methods in school. But at other times, I switched back to English because typing in English was much less time consuming. As part of my master's training, I learned the *Jyutping* system, a Cantonese Romanization scheme developed by a group of Cantonese linguists in Hong Kong. Since then, I have exchanged messages in Jyutping with my fellow linguistics friends who can also understand this system. Within our group, we see Jyutping as a unique system of communication that serves as some sort of "in-group" code among us. Example 1.1 is extracted from a personal MSN exchange about camera lenses between AL and me back in 2005.

Example 1.1 An MSN conversation

1 AL:	buy ng buy 17–85/@2xxx? (*Translation: Do you want to buy the 17–85mm lens for about 2,000 dollars?)*
2 Carmen:	hmm why?
3 Carmen:	whose?
4 Carmen:	我唔買舊野喎 (*Translation: I don't want to buy second-hand lenses.*)
5 AL:	ar Jo buy a 40D body only, but if we want to buy 17–85, then take out
6 AL:	new ar (*ar* is a Cantonese discourse particle)

In this short exchange between AL and me, a range of "codes" can be identified. For example, in lines 2 to 4, I move from using English in my questions "hmm why?" and "whose?" to making my stance in Cantonese represented in traditional Chinese characters. AL's response in lines 5 and 6 looks like English, but his messages also include some Cantonese words being spelt out, such as the particle *ar.* Note that Cantonese would have rarely been written out outside the online world, yet Cantonese web users have identified creative ways of representing their spoken language in digital communication. One of the aims of this book is to offer an understanding and explanation of complex multilingual online interactions such as this one.

When I was studying in England between 2004 and 2007, chatting on MSN was an indispensable tool of communication between me and my friends and family back home. A typical evening in my college room would involve writing my thesis on my computer in formal academic English and logging on to MSN and chatting with friends and family in an entirely different style of language. During that time, various social media platforms emerged. I started a blog to share stories about my life in the UK. Example 1.2 is a blog post about the progress of my thesis writing.

Example 1.2 A blog post with multilingual resources

I love formatting . . . Jun 27
- 42 figures
- 7 tables
- 74 extracts

Many people hate formatting.
But I think formatting is GR8, coz that's possibly the only thing that you can control in your thesis, and the only thing that makes your thesis look 'interesting' right now!
hmmm. . . . yes, I'm dak bit zai! *(Translation: Cantonese Romanization of 特別仔, a special person)*
12 days to go! hurray. . . .
Hg (abbreviation of hai6 gam2, "that's all for now")

Only a few of my close friends knew that this blog existed and could completely understand what I was talking about; I often inserted a line or two in our "secret" Jyutping codes. There were certainly issues of inclusion and exclusion of my audience (as discussed in Chapter 2). At the same time, I was aware that friends who could not read Chinese were also following my blog. So I still wrote mostly in English unless the blog post targeted only my Hong Kong friends.

In the past few years, my IM activity has moved entirely to the mobile phone, on which I regularly use WhatsApp, a mobile instant messenger, to stay connected with friends and family. Consistent with what I used to do on MSN, I still combine linguistic codes in my messages and I enjoy playing with *emoji*, a system of graphic symbols and emoticons. I have been a Facebook user since 2007, and now it is one of my most visited social network sites; I regularly read and send Facebook posts from not only my desktop computer but also from my smartphone and tablet devices. I have two Facebook accounts: One for my close friends and family and another for my students and colleagues. In my work Facebook, I post mainly to my course "groups" to interact with my students. I deliberately write in English only when interacting with students (although I sometimes add emoticons), as the medium of instruction of my courses is English; whereas on my personal Facebook wall, I draw on a wider range of languages, scripts, and modes, depending on my audience and the content of the post. I am also a regular user of other digital media such as Flickr, Google Scholar, Instagram, Pinterest, YouTube, and Wikipedia, where I constantly come across texts that are multilingual, multiscriptual, and multimodal. For example, on Flickr, I alternate between Chinese, English, and Chinese-English mixed code when it comes to writing captions, tags, and comments. For information searches on Google and Google Scholar, I use mostly English keywords for my academic work, but at other times I input search queries in Chinese only.

The previous narrative of my technology-related life, or my auto-technobiography (Kennedy, 2003; Barton and Lee, 2013), reveals what is actually happening to language(s) and written texts in the age of the internet and is summarized as follows:

- Language is an indispensable element in online communication. All of the platforms I have mentioned in my auto-technobiography rely heavily on the written word, even though words are often combined with other modes. The centrality of language in the digital age has been discussed in greater detail in Barton and Lee (2013).
- Texts are constantly produced and read over the internet by users in different physical locations. These texts, as in Examples 1.1 and 1.2, may contain multiple linguistic resources including different scripts and languages.
- Language choice on the web does not always reflect language use in offline communication contexts. As technological affordances change, our linguistic practices respond to these changes. My brothers' and my own preference for English online in the early 1990s was not simply a matter of choice, but our response to the technological constraints during that time. In offline contexts, we used very little English, except for lessons in school. This early tendency to use more English on the internet also echoes the widespread discourse of the global status of English in the 1990s (see discussion in Chapter 2). As we started to come across newer affordances or possibilities for meaning-making online, we were able to make decisions about our ways of writing online. Code choice and code-switching have become salient themes in research on online communication (see Chapters 2 and 3).
- New affordances and possibilities offered by digital media give rise to creativity and identity performance online (Chapter 4). As shown in my auto-technobiography, my deployment of languages over my two Facebook accounts allows me to juggle between the various roles I play in life, such as a friend, a family member, a teacher, and so on. The new possibilities offered by digital media also foster new forms of interaction across the globe. For example, globalized social media such as Flickr and YouTube provide translocal interactional spaces for people from around the world to form online communities. In these translocal spaces, it is not uncommon for multilingual web users to talk about the languages they know and how they are used online (see Chapter 5). With the help of online translators and other tools, new forms of multilingual encounters are made possible online (Chapter 7).
- Languages that used to have no standard script or minority languages that were not represented in writing are now made more visible in online communication (see Chapter 6). In my case, Cantonese is essentially a spoken language; before the digital age, authentic Cantonese conversations had never been recorded in writing except for deliberate productions of transcripts (e.g. witness testimonies in law courts). In online writing spaces, nonetheless, people have found ways of representing spoken Cantonese in writing. The use of colloquial Cantonese writing is also rising in other domains as a reflection of its widespread use in internet communication.
- What I do online is tied closely to my offline lived experiences. For instance, Example 1.1 is a seemingly mundane shopping-related discussion, and the blog post in Example 1.2 is my written reflection on the progress of my thesis writing. This also means that research into language on the internet must take into account not only texts on the screen but also what people do with these

texts in other areas of their lives. (See Chapter 8 for a more detailed discussion of research methods.)

For the time being, the singular form *language* refers generally to any system of communication, whereas the plural *languages* are traditional categories used to refer to distinctive systems of linguistic codes defined by groups of speakers, though I am aware that both notions have been challenged in recent sociolinguistic research, as I explain in the section 'Beyond multilingualism' in this chapter.

I have started this book with my own story. Some readers may identify with my experiences, while others may do things differently on the web. This book is not just about how people deploy languages online, but also about the broader social practices that concern hundreds of millions of people around the globe. In the rest of this chapter, I explain why multilingualism is central to the growing field of internet linguistics and the theoretical approach taken throughout the book.

WHY MULTILINGUALISM ONLINE?

The internet was a predominantly English medium, both in terms of users and contents, especially in the 1990s (see Chapter 2). This partly explains why early research on computer-mediated communication (CMC) focused largely on linguistic features based on English. However, predictions about the dominance of English on the web were soon challenged by scholars (Crystal, 2006; Danet and Herring, 2007) and numerous market surveys (e.g. W3techs and Internet World Stats) which provide further evidence for the presence of languages other than English on the internet. Since the late 1990s, access to the internet has become more affordable, which has given rise to the growth of mass communication across geographical borders. This is partly owing to the rapid advancement of technological affordances. For example, inputting and processing non-Roman scripts on the computer have been made much easier. Like me, many web users were gradually becoming more at ease when drawing on multiple linguistic resources, or codes (more on this terminology later), when participating in online activities.

According to Internet World Stats (2015), web users from Asia already outnumbered users in other parts of the world. Although measuring internet users by geographical locations does not accurately represent the level of linguistic diversity on the web, the figures are still indicative of the possible range of non-English languages available on the web. Statistics from Internet World Stats (Table 1.1) also reveal the large proportion of non-English language speakers on the web, with Chinese, Spanish, and Arabic being the larger ones after English (Internet World Stats, 2015).

In response to the growing diversity of users and their language resources, some CMC researchers have turned their attention to non-English interaction on the internet. Notably, in 2007, exactly a decade before the publication of this book, the first comprehensive edited volume devoted to multilingualism on the internet was published under the title *The Multilingual Internet* (Danet and Herring, 2007). The coverage of the 18 chapters is remarkable in terms of languages, online platforms, and linguistic topics. Most of the studies reported in the volume

Table 1.1 Top 10 languages used on the web (adapted from Internet World Stats, 2015, www.internetworldstats.com/stats7.htm)

Top 10 Languages on the Internet	*Internet Users by Language*	*Internet Penetration (% Population)*	*Users Growth in Internet (2000–2015)*	*Internet Users Percentage of World Total (Participation)*
English	872,950,266	62.4%	520.2%	25.9%
Chinese	704,484,396	50.4%	2,080.9%	20.9%
Spanish	256,787,878	58.2%	1,312.4%	7.6%
Arabic	168,176,008	44.8%	6,592.5%	5.0%
Portuguese	131,903,391	50.1%	1,641.1%	3.9%
Japanese	114,963,827	90.6%	144.2%	3.4%
Russian	103,147,691	70.5%	3,227.3%	3.1%
Malay	98,915,747	34.5%	1,626.3%	2.9%
French	97,729,532	25.4%	714.9%	2.9%
German	83,738,911	87.8%	204.3%	2.5%
Rest of the languages	734,013,009	31.3%	1,042.9%	21.8%

were carried out in the contexts of the contributors' native languages, including Arabic, Chinese, French, Greek, Japanese, Spanish, and Thai. Data analyses in these studies were mostly based upon existing methods and concepts in linguistics and sociolinguistics, such as descriptions of linguistic features, language choice, code-mixing, gender, and politeness, some of which are also explored in this book. Although earlier publications in other languages had already reported multilingual resources in CMC prior to this volume (e.g. Anis, 1999, in French; Androutsopoulos and Schmidt, 2002, in German), *The Multilingual Internet* has provided solid theoretical and methodological foundations for subsequent work in the area of multilingualism online. Since the publication of *The Multilingual Internet*, the body of studies on multilingualism online has grown significantly. One of the main purposes of this book is to bring together these important works, old and new, as well as my own research carried out in Hong Kong over the last 10 years.

One of the most important reasons why linguists should pay attention to multilingualism on the internet is that, as Crystal (2006: 229) puts it, the web "offers a home to all languages". The huge database of freely and often publicly available multilingual texts online becomes a rich source of authentic written data for researchers across disciplines, including linguistics, discourse analysis, media studies, and social sciences. In fact, except for transcripts deliberately produced for spoken discourse, the web is by far the only space where ordinary people's written conversations are performed (in real time) and archived. As Sebba (2012) notes, written communication has long been underrepresented in research on multilingualism and code-switching. There is a pressing need for researchers on multilingualism to take into account this growing source of written data mediated by virtual linguistic landscapes.

Recent years have seen a major attention shift to analyzing language data on the so-called Web 2.0 media such as blogs, Facebook, and Twitter. With the advent of social media, the web is no longer just about connecting information and hypertexts, but also about connecting *people* (O'Reilly, 2005). The affordances of self-generated contents, easy self-publishing, and participatory culture in social media further strengthen what Manuel Castells refers to as the *network society*, which is

> a social structure based on networks operated by information and communication technologies based in microelectronics and digital computer networks that generate, process, and distribute information on the basis of the knowledge accumulated in the nodes of the networks.
>
> (Castells, 2006: 7)

Castells's work on the *network society* is especially relevant in conceptualizing changes to linguistic practices online. He points out that unlike earlier forms of online communication tools, such as instant messaging and email, Web 2.0 technologies including blogs, Facebook, and YouTube give rise to a different form of online interaction, or what he calls "mass self-communication" (Castells, 2010: xxvii), in which self-generated contents (which can be words, images, videos, or a combination of all these) are often made available to large networks of self-selected or public audiences. This has drastically transformed the infrastructure of the already existing network society which used to be shaped by more private forms of social interaction, such as telecommunication or even email. Herring (2011) coined the term *convergent media computer-mediated communication* (CMCMC) to characterize the changing structure of discourse in convergent media. For example, YouTube videos often coexist with written comments from viewers around the world who may not share the same language(s) with the original poster of the video. For linguists, such mass global production and circulation of texts online provides an important source of informal, interactive, multimodal, and easy-to-access linguistic data for investigation.

Crucially, the network society operates within "spaces of flows", which Castells (1999: 295) defines as the "material arrangements that allow for simultaneity of social practices without territorial contiguity". The web is one of many components of such spaces of flows. Other forms of flows may exist simultaneously outside the online space. For one thing, we are living in an increasingly mobile world. Flows of people, knowledge, ideas, and objects are all speeding up, leading to new multilingual encounters between people across spaces, online and offline. These ideas are also captured in Vertovec's (2010) term *super-diversity*, originally proposed to understand the increasingly complex patterns of mass migrations. Vertovec (2010: 83) notes that

> more people are now moving from more places, through more places, to more places. . . . Today newer, smaller, transient, more socially stratified, less organised

> and more legally differentiated immigrant groups comprise global migration flows. . . . Super-diversity is a term intended to capture a level and kind of complexity surpassing anything many migrant-receiving countries have previously experienced.

All these have important implications for language use and multilingualism. First of all, such flows of people and information across the globe naturally result in flows of languages as well. Production and transmissions of new forms of texts are made possible through creative deployments of multimodal meaning-making resources, with support from new technological affordances and tools such as fast-paced wireless internet, smartphones, open source software and applications, Creative Commons, and so forth. A major linguistic phenomenon that global spaces of flows have introduced is the increasing use of local (and minority) languages among diasporic communities in online communication, a topic that is pursued in Chapter 6.

The idea of superdiversity has also been taken up by sociolinguists (e.g. Creese and Blackledge, 2010; Blommaert, 2011; Blommaert and Rampton, 2011), who have begun to question the validity of traditional sociolinguistic concepts such as code-switching and speech community in this super-mobile and diverse world. This is pursued in detail later. Undoubtedly, sociolinguistic theories and methods need to be revisited to reflect the changes brought about by increasingly complex social networks. One needs to move beyond a descriptive approach to multilingualism, which has long focused on identifying distribution of languages and structural patterns of code-switching. While these traditional concepts still serve their purpose, they have been largely developed and applied in analyses of spoken data in non-CMC contexts, especially in education. In response to some of these changes, Androutsopoulos (2013a), also drawing on Castells's idea of the network society, coined the term *networked multilingualism* as a way of understanding people's multilingual practices on social network sites. In the next section, I show the ways in which sociolinguists are taking up the concept of superdiversity to rethink the meaning of multilingualism in the 21st century. In particular, I provide an overview of alternative terms to multilingualism and explain how and why the term *multilingualism* is still adopted in this book.

BEYOND MULTILINGUALISM

The title of the book, *Multilingualism Online*, needs explaining at this point. In addition to providing a working definition for the book's title, the primary aim of this section is to demonstrate the extent to which my understanding of multilingualism relates to recent developments in bi- and multilingualism research.

- *Multilingualism*: As an established concept in sociolinguistics, *multilingualism* broadly refers to the coexistence of two or more languages. Sometimes used interchangeably with *bilingualism*, the term is often used in the literature to describe either a person's knowledge in two or more languages (e.g. a multilingual speaker) or a social situation where two or more languages are present (e.g. a multilingual society). Throughout the book, *multilingualism* is used as a cover term when talking about the use and production of texts in any situation where two or more languages are involved either in the text per se or are a theme in the text, regardless of degrees of speakers' proficiency or power relations between the languages concerned.

- *Online*: The title *Multilingualism Online* deliberately borrows the structure of the title of my earlier co-authored work, *Language Online* (Barton and Lee, 2013). It is made clear in *Language Online* that the book "does not assume a strict online-offline dichotomy" (7), nor does it view digital linguistic practices as exclusively online. In the present work, I continue to understand *online* as "the broad context and domain" of research where interactions take place via networked devices such as personal computers and mobile phones. At the same time, I take it that practices that may be seen as exclusively online phenomena may shift to offline linguistic landscapes, as illustrated in Lee (2015). Other terms may be used throughout the book to refer to specific online spaces or contexts, such as *computer-mediated communication (CMC), online media, digital media*, and *digital communication.*

The umbrella term *multilingualism* and its associated concepts such as code-switching and code-mixing have been challenged in a number of recent approaches to bi- or multilingualism. Specifically, traditional understandings of multilingualism have been criticized for being misleading or for failing to capture changes in meaning-making processes in the age of superdiversity. Alternative terms have thus been proposed. Notably, these include *translanguaging* (García, 2009; García and Li, 2014), *translingual practices* and *code-meshing* (Canagarajah, 2011, 2013), *polylingual languaging* (Jørgensen, 2008), and *metrolingualism* (Otsuji and Pennycook, 2010). These competing yet interrelated terms are difficult to define. Some of these have also been applied to digital discourse research (e.g. Benson, 2015; Hafner et al., 2015). Regardless of their intended distinctions, all these terms and their associated critiques of multilingualism share at least some of the following arguments and insights into sociolinguistics:

- Languages are not static categories, but ideological constructs. There are no clear-cut boundaries between languages (see Heller, 2010).
- Multilingualism is not about using one language at a time, nor is it just about the different languages serving different functions at different times (see Cenoz and Gorter, 2011). Instead of seeing the multiple languages as independent linguistic systems where one adds on to another (often referred to as *additive bilingualism*), these new terms see relationships between languages as complementary, nonadditive, and fluid (García, 2009).
- Research on multilingualism should shift its focus from speakers' linguistic competence to what they do with their linguistic resources in everyday social interactions. Language users regularly draw on a wide range of multimodal meaning-making resources regardless of their knowledge in the languages concerned. Canagarajah (2011) proposes the term *code-meshing* to describe

the multiplicity of symbols and modes in people's translingual practices (see Martin-Jones and Jones, 2001; Sebba, 2012).

- When codes or languages coexist in an utterance or a social interaction, the relationships between the languages are not stable. Instead of identifying a language being the matrix code to which another code is added (Myers-Scotton, 1992), it is more meaningful to analyze the multilingual textual data at hand holistically. Code-switching can be seen as a resource for language users to claim authenticities and assert identities (Pavlenko and Blackledge, 2004; Blackledge and Creese, 2013; see also Chapter 4).

These are important issues that traditional understanding of multilingualism needs to address. While I share the view that languages are mobile resources and that relationships between languages are complex and dynamic, labelling individual languages is still meaningful in that, as we will see in some of the data in the book, ordinary language users often talk about their meaning-making resources in terms of individual *languages*; it is necessary to use socially accepted boundaries between languages to understand people's cultural knowledge related to the languages they use online. In fact, for practical reasons, none of the newly proposed approaches outlined previously can completely abandon *languages* and *multilingualism* altogether. From time to time, these words are brought up in discussions of translanguaging, translingual practices, and the like. As Canagarajah (2013: 15–16) argues:

> while language resources are mobile, they acquire labels and identities through situated uses in particular contexts and get reified through language ideologies. Therefore, labeled languages and language varieties have a reality for social groups. More importantly, they are an important form of identity for these groups.

In addition to languages, related units such as *codes*, as in code-switching and code-mixing, as well as *resources* are adopted in this book where appropriate. In online interaction, meaning-making resources are not limited to linguistic codes; they also include other semiotic modes of communication including images and videos. In this book, I define *resources* as the materials and knowledge available to text users to draw upon in text- and meaning-making. Resources do not simply exist inherently and biologically in human beings, nor are they obligatory. Rather, they are socially shaped and constructed by a number of factors such as experience, beliefs, values, and social backgrounds of the participants. (See also Ivanič and Tseng, 2005 for a discussion of the concept of resource.)

Overall, *multilingualism* is still a useful umbrella term to refer to the use and co-occurrence of multiple linguistic resources in any context. In research on CMC, the term has been used to refer to various forms of linguistic phenomenon online (Androutsopoulos, 2013b: 671):

- *Linguistic diversity on the internet*: This kind of research considers the internet as a whole multilingual space and measures the distribution of languages online by content or the first language of internet users (see discussion in Chapter 2).

- *The co-existence of languages on a webpage/website*: The various linguistic codes only co-occur on a website but may not be related dialogically. For example, on Flickr, a user may configure their site to the English interface while writing captions in Chinese only. Here, the two languages coexist on the same page, but their functions are separated. Another example would be the emblematic use of certain languages to represent identity groups, such as using multiple scripts in the logo of a website.
- *Multilingual users using different languages in their online interactions*: This category of multilingualism is concerned with the ways in which users deploy their multiple linguistic resources in asynchronous and synchronous online interactions (see Chapters 2 and 3). These may include a thread of YouTube comments in different languages written by people who do not share a common language or a one-on-one chatroom conversation in which the writers alternate between languages they both know.

Without deliberately adopting new terms, my use of multilingualism here can still be extended to include many of the observations and arguments made by the emerging approaches outlined previously. For a long time, "multilingual CMC" researchers have already moved beyond studying linguistic diversity online or simply "counting languages". Numerous studies have examined how web users draw on and play with linguistic and other semiotic resources to achieve their own purposes and to perform identities online (e.g. Androutsopoulos, 2006; Koutsogiannis and Mitsikopoulou, 2007; Lam, 2009; Tagg and Seargeant, 2012). My own research on IM and Facebook has focused largely on the relationship between language choice and people's situated practices (Lee, 2007a, 2011, 2014). These studies also respond to the call for a third wave of internet linguistic research in Androutsopoulos (2006, 2008), which accentuates language users' situated practices and identities instead of just describing conversational or linguistic features of log data (as in earlier waves of CMC research). It is also worth noting that one need not possess knowledge of multiple languages or literally be called a bilingual person to engage in multilingual activities on the internet. On the photo-sharing site Flickr, for example, it is possible for someone who has limited knowledge of English to write captions or tag a photo in English only or to engage in a discussion about different languages entirely in English. Such talk about languages is a manifestation of multilingualism in which people participate in multilingual activities through metalinguistic discourses about languages (see Chapter 5).

A PRACTICE-BASED APPROACH TO MULTILINGUALISM ONLINE

It becomes clear from the earlier discussion that sociolinguists are moving beyond describing language varieties to making a stronger link between language and social life. The internet gives rise to new social practices. At the same time, social practices are being transformed as communication moves online. Much of this transformation is mediated by language and the reading and writing of texts. Some examples of people's lived experiences involving digital text-making include sharing holiday photos with friends on Facebook, discussing news and politics or even organizing social movements in social media such as Twitter, learning a new language by watching videos on YouTube, and so on. In these globalized and superdiverse online spaces, multilingual encounters are becoming a norm, regardless of how many languages one can speak.

Reflecting on my own online practices, as in the auto-technobiography presented earlier in the chapter, allows me to understand better the complex relationship between my lived experiences and my deployment of linguistic resources online. In researching multilingualism online, then, it makes sense to focus on *people*, that is, language users and their practices. In fact, the word *practice* is frequently mentioned in the alternative approaches to multilingualism outlined in the previous section, although few have given it a clear definition. My understanding of practice comes from *literacy practices* in the field of *literacy studies*, a mid-level sociocultural theory of reading and writing which posits that literacy is best described in terms of people's literacy practices (Barton, 1994, ; Barton and Hamilton 1998,). *Literacy practices* is a central notion in literacy studies and it generally refers to the common patterns in reading and writing in a context in which people bring in their cultural knowledge to an activity (see Barton and Lee, 2013: 24–25). As outlined in the previous section, many of the recent arguments about multilingualism and sociolinguistics broadly share some of the viewpoints in literacy studies. My aim in this book is to bring together traditions of literacy practices and recent understandings of sociolinguistics in conceptualizing multilingualism online. Multilingualism online, in this book, is examined in terms of *multilingual practices*, which are "a subset of literacy practices that involve people doing things with two or more languages" (Barton and Lee, 2013: 44). In sum, multilingual practices are literacy practices embedded in people's everyday activities in which multiple linguistic resources play a role. And this book pays specific attention to multilingual practices that are mediated by online communication.

Another central notion to multilingualism online is *text*, as texts are "part of the fabric of social life" (Barton and Lee, 2013: 11). Texts are not only a product of language but are mobile social constructs, and they shape and are shaped by social practices. In the online world, texts are constantly produced and published and get read, shared, and circulated across the globe. This means that in addition to paying attention to people's practices, observing the texts they use and produce is equally important in researching multilingual practices. This analytical approach that involves both texts and practices is very much in line with what Androutsopoulos (2008) refers to as *discourse-centered online ethnography* (DCOE). In doing DCOE of the multilingual internet, the researcher moves back and forth between observing patterns of discourse on the research site and talking to language users (e.g. through interviews and collecting technobiographies) to elicit their insider perspectives about their language use. As we see in the book, while acknowledging the importance of understanding what we can observe in textual data, researchers should closely examine what people do with their resources in online communication and how they make sense of what they do. (See Chapter 8 for a more detailed discussion of methods.) As some of the data in the book illustrate, people do not always talk about their text-making resources in the same way, thus revealing individual differences in practices.

OVERVIEW OF CHAPTERS

In sum, the book aims to bring together traditional and up-and-coming research on the topic of multilingualism online, providing an updated overview of the topic that brings together the established area of multilingualism in sociolinguistics and the emerging area of online communication. Online platforms covered in the book range from traditional forms such as email, chatrooms, and IM through to social media including Facebook, Flickr, Instagram, Twitter, WhatsApp, and YouTube.

The chapters that follow take a progressive approach to multilingualism online, from micro-level analyses of linguistic resources to broader issues of metalinguistic discourse and identities. Chapters 2 and 3 start with identifying resources available to online users for meaning-making and describing patterns of language choice and code-switching, two of the prevailing topics in existing research on multilingualism online. In response to recent developments in multilingualism online, the book then gradually moves beyond descriptions of code choice to understanding what multilingual practices mean to web users, how multilingual practices change throughout people's technology-related lives, and how multiple layers of discursive resources are deployed, mixed, remixed, recontextualized, or played with to assert identities or to achieve specific goals.

Chapter 4 looks into the ways in which multilingual practices become powerful resources for identity construction and performance online. The book highlights the importance of understanding multilingualism from a social practice point of view, that is, looking at details of what people do with languages and how they talk about languages online. People's metalinguistic representations of their own and others' multilingual practices are detailed in Chapter 5. The discussion of multilingualism in this book also demonstrates that, in the age of social media and superdiversity, traditional linguistic boundaries which used to be defined by regions and speech communities, or even dichotomies such as online–offline and global–local, begin to break down. Chapter 6 outlines the ways in which some of the so-called minority languages and languages that do not have standardized writing systems develop into being used in online communication. Chapter 7 examines the role of translation in the age of "glocal" communication. Topics related to translation online range from the policy governing multilingual Wikipedia entries to the use of free tools such as online dictionaries and built-in translators on Facebook. The chapter also examines how practices of folk translation impact people's everyday lives. Chapter 8, the closing chapter, offers a discussion of the methods and challenges in researching multilingualism online. In particular, it draws attention to the researcher's roles and positions. Finally, it points to a few directions for future research.

2

LINGUISTIC DIVERSITY AND LANGUAGE CHOICE ONLINE

Overview

- The internet as an English medium
- Measuring linguistic diversity on the internet
- Language choice and multilingual resources online
- When linguistic resources meet other semiotic modes in CMC
- Understanding the affordances of meaning-making resources
- Beyond counting languages

The internet was first developed in English in the United States. A major debate within the literature of multilingualism online in the 1990s was whether the spread of the internet would bring about new relations between English and other languages of the world. Concerns were also expressed as to whether English would "rule" the internet. This partly grew out of intense scholarly interest in the topic of language and globalization. Academic responses to this inquiry were twofold: Some believed that the internet encouraged the growth of English, while others argued that the spread of a global network would give rise to linguistic diversity. The primary aim of this chapter is to map out the linguascape of the internet (Ivković and Lotherington, 2009; Thorne and Ivković, 2015) and to provide an overview of the relationship between English and other languages on the web.

Drawing on older and more recent findings and arguments, this chapter begins with a discussion of the status of English on the internet. It then presents evidence of the presence and rise of other languages online through looking into results from market surveys and quantitative academic research. Alongside these approaches, qualitative case studies of language choice on various online platforms are presented to illustrate the ways in which resources are drawn upon on the internet. In addition to identifying resources, the chapter also points out that language choice online is not simply determined by technologies, but is often negotiated within a range of ecological factors.

THE INTERNET AS AN ENGLISH MEDIUM

> American-English is quite clearly the language of the internet.
>
> (Abbot, 1998: 92)

> Nowhere is the effect of this expansion of English into new domains seen more clearly than in communication on the Internet and the development of "net English".
>
> (Graddol, 1997: 2)

> The dominance of English in the Internet needs no arguing for. Computers are in any case English-oriented.
>
> (Gupta, 1997)

> If you want to take full advantage of the Internet there is only one real way to do it: learn English.
>
> (Specter, 1996: 1)

These quotes, all taken from publications in the 1990s, demonstrate the strength of both academic and popular beliefs about the possible dominance of English on the internet. Various reasons may have contributed to this tendency of seeing English as the primary language online. First, looking through the history of the internet, it is not surprising that the internet was very much English-oriented. The internet and its related technologies began and became popularized in parts of the world where English is spoken as a main language. The internet, or the ARPANET (an earlier prototype of the present day internet), came to first use in the US while the World Wide Web was invented by the work of the English scientist Tim Berners Lee. This naturally resulted in early technological innovations and information exchanged over the net being designed by and written for native English speakers. According to the Worldmapper (2006), a joint project of mapping the world's data by researchers at the University of Sheffield and University of Michigan, most internet users in 1990 resided in North America, amounting up to 70% of the total internet population. Second, against the backdrop of globalization and discourses of the "hyperglobalizers" (Dewey, 2007), who believe in the "denationalization" of economies, is the idea of English being *the* global language (Kachru, 1992; see also discussion in Crystal, 1997). A third perspective that gave rise to the emergence of a global language on the internet in the 1990s has much to do with the limits of technological affordances. For one thing, the design of the QWERTY keyboard, which is still commonly used today, favors English and other alphanumeric scripts (Danet, 2010); the inputting or displaying of nonalphabetic symbols was not fully supported on personal computers in the early 1990s. It is also likely that these are the reasons why some early research on CMC tended to focus on English texts only. As Danet and Herring (2007: 5) observe, "most researchers publishing in English venues have generalized about the language of computer-mediated communication, whereas in fact they were describing computer-mediated *English*".

At the same time, linguists also made use of quantitative surveys in support for the possible global status of English (Crystal 1997; Fishman 1998). One of the most cited figures is found in Fishman (1998), which reports that over 80% of internet content was written in English and about 90% of the world's internet servers were based in countries where English is used as the primary language. Fishman further predicted that "English will likely remain the single most commonly used language on the Internet for the foreseeable future" (1998: 34). Similarly, Babel (1997, cited in Crystal, 2006), a survey of the content languages of over 3,000 websites, revealed that a high percentage (82.3%) of webpages were in English. In addition, the latest figures from W3techs (2015) show that English still accounts for the largest percentage (55%) of the languages recorded on the world's most popular websites. As a result, concerns were also expressed as to whether the growth of English on the internet would threaten smaller languages and lead to *linguistic imperialism*, in which "the dominance of English is asserted and maintained by the establishment and continuous reconstitution of structural and cultural inequalities between English and other languages" (Phillipson, 1992: 47).

There are concerns about the potential threat of English to smaller languages or endangered languages even in today's age of digital social media. The popular search engine Google was found to favor larger languages. Although over 150 interface languages are available on Google, it can only recognize search words and produce results in many fewer languages (Prado, 2012). Pann and Phatak (2012) found that only 20% of the local search engines developed by Google were devoted to Asian languages, while Prado (2012) reported that only one African language could be recognized by Google, and indigenous American languages were excluded. The possible dominance of English is also evident on the microblogging site Twitter. Tracking languages of tweets over four periods of time on a day, Honeycutt and Herring (2009) found that English was the most dominant language of the four sets of tweets. Later, Hong et al.'s (2011) large-scale survey of linguistic behavior on Twitter reported that over half of the tweets were written in English, while none of the other languages counts toward more than 10% of the tweets collected.

Global media companies are aware of the increasingly multilingual online population and the potential spread of their user base to people who do not use English as their everyday language. This is evidenced by translations of web interfaces and contents, such as the different language editions of Wikipedia, and built-in translation functions such as the "see translation" feature on Facebook. Such attempts to diversify the web are not only restricted to languages per se. In April 2015, Apple updated its iPhone pre-installed emoticons (emojis) to include faces of different skin colors and added national flags of more countries. Because iPhone can only support inputting in a limited number of languages, this move can be interpreted as the company's attempt to index global diversity through more "universal" visual symbols. Whether these strategies can really address multilingualism and diversity is still questionable. For example, the multilingual status of Wikipedia is a highly complex one. To date, the site provides over 280 language editions. The different language editions are not literal translations of the same set of articles; each language edition is an independent site with its own homepage and articles written in the same language. Among the 280 language editions, the English edition still has the largest number of articles (Hecht and Gergle, 2010), and similar to the case of Google, many Asian and African languages are still absent (Graham, 2012).

Smaller languages do exist on Wikipedia. However, with insufficient first-language content writers and "language police", the actual language used often deviates from its conventional usage. The isiXhosa edition is an example in point. IsiXhosa is one of the official languages of South Africa, spoken by approximately 8 million people. As of July 2016, only 533 articles are available on the isiXhosa Wikipedia. Most of the article authors, as Deumert (2014) notes, are "language enthusiasts" from the US, Europe, and Australia who have admitted to their limited knowledge of isiXhosa. As a result, some of the texts on the isiXhosa Wikipedia are only written in a "simulated" form of unsystematic isiXhosa; this is not real representation of the language. And unlike the English articles, which are heavily policed and constantly edited, the erroneous isiXhosa articles would remain uncorrected for the time being (Deumert, 2014). On the other hand, the English edition of Wikipedia may only be predominant to a certain extent. Many topics are found to be language-specific and are available only in a language other than English. Hecht and Gergle (2010) conducted a comparative study between the English and the German Wikipedia editions and found that although there are still three times as many English articles as German ones, they cover only half of the concepts in the German edition. It then becomes clear that the distribution of languages online cannot be measured through one single criterion and approach. For example, Herring et al. (2007) made use of multiple analytical methods, including random analysis of 1,000 journals and language network analysis, to measure the level of linguistic diversity of LiveJournal.com, one of the earliest blogging services. Their results suggest that although English still dominates on the site at the global level, it coexists with a growing number of non-English monolingual blog entries at the local level. Other patterns may emerge as other factors are taken into account, such as the breakdown of content types (as in the case of Wikipedia) and the actual linguistic behavior of users, such as whether a post on Twitter is a retweet. Websites may also automatically select contents in a specific language for users once they have set a default language.

Taken together, the different sources of data and findings about online linguistic diversity suggest that, on the one hand, uses of English continue to grow on the internet; on the other hand, other languages are on the rise. As will be detailed in the next section, for different reasons, many languages have been growing at a much faster pace than English over the years.

MEASURING LINGUISTIC DIVERSITY ON THE INTERNET

While the status of English on the internet may be uncertain, more recent public surveys and quantitative research have revealed drastic changes in the distribution of English and other languages online. Statistical findings are generated either through recording the main language of individual webpages or websites or by identifying the native language of web users. Some of the most cited public surveys include Internet World Stats (www.internetworldstats.com) and W3Techs (http://w3techs.com), both of which are hosted by marketing companies. As of 2015, over 70% of internet users in the world are speakers of languages other than English (Internet World Stats, 2015). Several Asian regions, including Hong Kong, Japan, South Korea, and Taiwan, are reported to be amongst the top 50 regions with the highest population of internet users (Internet World Stats, 2015). In some of these areas, English is rarely used in people's everyday lives outside the internet.

Table 2.1 is a side-by-side listing of languages of the world in 2015 by content languages of websites (W3techs, 2015), number of internet users (Liao, 2015a), and number of native speakers measured offline (Ethnologue, 2015).

Table 2.1 Top 20 world languages online and offline

Rank	*Language (Content of Websites)*[i]	*Websites Percentage (Million)*	*Language (Internet Users)*[ii]	*Internet Population (Million)*	*Language (Offline)*[iii]	*Native Speakers (Million)*
1	English	553	English	600.9	Chinese	1,197
2	Russian	58	Chinese (simplified)	564.2	Spanish	399
3	German	57	Spanish	217.5	English	335
4	Japanese	50	Arabic	110.9	Hindi	260
5	Spanish, Castilian	45	Portuguese	104.6	Arabic	242
6	French	39	Japanese	104.6	Portuguese	203
7	Chinese	27	Russian	104.2	Bengali	189
8	Portuguese	25	French	80.2	Russian	166
9	Italian	20	German	78.1	Japanese	128
10	Polish	18	Hindi	77.7	Lahnda	88.7
11	Turkish	16	Korean	44.9	Javanese	84.3
12	Dutch, Flemish	14	Wu Chinese	37.4	German	78.1
13	Persian	9	Italian	37.2	Korean	77.2
14	Arabic	7	Turkish	35.6	French	75.9
15	Korean	7	Vietnamese	34.8	Telugu	74.0

(Continued)

Table 2.1 (Continued)

Rank	*Language (Content of Websites)*[i]	*Websites Percentage (Million)*	*Language (Internet Users)*[ii]	*Internet Population (Million)*	*Language (Offline)*[iii]	*Native Speakers (Million)*
16	Czech	7	Chinese (traditional)	30.2	Marathi	71.8
17	Swedish	5	Urdu	28.3	Turkish	70.9
18	Vietnamese	5	Egyptian Arabic	26.6	Tamil	68.8
19	Indonesian	4	Bengali	25.6	Vietnamese	67.8
20	Greek	4	Indonesian	25.4	Urdu	64.0

[i] W3techs (2015): http://w3techs.com/technologies/overview/content_language/all
[ii] Liao (2015a): http://people.oii.ox.ac.uk/hanteng/2015/02/24/internet-users-by-language-the-top-20/
[iii] Ethnologue (2015): www.ethnologue.com

The information in the three columns in Table 2.1 needs some further interpretation. The left-most column contains statistics of "usage of content languages for websites" provided by W3techs (2015). It measures the content language of the world's 10 million most-popular global websites. As can be seen, English is the major language on only 55.3% of these. This is a significant decrease from the 80% reported by Fishman in 1998. Interestingly, the figures from W3techs also suggest that languages that have relatively small number of speakers in the offline world, such as Czech and Swedish, are amongst the languages with stronger online presence.

The second and third columns provide some points of reference for the correlation between number of native speakers and internet penetration. Some of the most spoken languages of the world, such as Bengali and Hindi, which are amongst the top 10 largest languages with high internet penetration rate (Ethnologue, 2015), have less than 0.1% contents on the web according to Liao (2015a). This can be explained by a number of factors. For the case of Indian languages, although there have been an increasing number of websites in Indian languages, the nonstandard character encoding systems used by these languages are not readily recognizable by major search engines (Majumder et al., 2006). Another reason is that W3techs only measures languages on popular "global" sites that are based in the US and China. These global surveys have neglected the increasing number of web-based publications in Indian languages (news, blogs, magazines) used primarily by local communities. More efforts need to be made to measure a more comprehensive range of criteria and local identifiers for more accurate analysis of linguistic diversity online (Liao, 2015b).

Despite researchers' constant efforts to measure the level of linguistic diversity on the internet, no single method developed to date can accurately represent the linguistic situation on the internet. For example, Internet World Stats (2015) admits that in their surveys, "many people are bilingual or multilingual, but here we assign only one language per person in order to have all the language totals add up to the total world population". The methodologies adopted in these marketing surveys are certainly questionable, as Paolillo (2007) also notes. First, the estimates generated for each language are often based on different sources from different countries that do not adopt the same methodology to measure internet penetration rates. Second, many of the figures are provided by governments (e.g. national censuses) and marketing companies who may have reported sources in favor of their own interests, which may then lead to biased results or even overestimates. Third, some smaller language groups may not be surveyed as regularly as the larger languages. In light of these problems, Paolillo (2007) proposed a new method of analyzing linguistic diversity. This method, or what Paolillo refers to as the *linguistic diversity index*, draws on data from not only one single representative language, but varieties of languages (or "language groups") within a region or a country. For example, the US has 170 language groups with a diversity index of only 0.78, while Africa, which has 2,390 language groups, scores 185.68 in linguistic diversity. With this method, Paolillo reveals that regions that are less linguistically diverse are more likely to get fuller access to the internet. This is because the languages associated with these regions need less technological solutions compared to those with a huge range of language varieties and scripts. Based on his findings and observations from existing surveys, Paolillo concluded that internet technologies still favor larger languages.

A further drawback of most quantitative studies of linguistic diversity is that they often assume that an individual web user speaks only one language or uses one language at a time, and they fail to sufficiently document the array of linguistic resources people use for actual instances of meaning-making online. Clearly, the figures presented here must be treated with caution. Nonetheless, they still provide indications that the presence of languages other than English on the web is on the rise. In countries beyond North America and Europe, access to the internet has become much more widespread and affordable. Typing in non-Roman scripts on the computer is now possible and relatively easy, and most foreign character systems are readable on major computer platforms and web browsers (though not necessarily recognizable by search engines). These improved technological affordances provide web users, especially those who do not use English as a primary language, the opportunities to communicate in their local languages online. Dor (2004) also argues that as more people who do not speak English as their first language have access to the internet, there is a greater demand for localizing online contents such as translating the web into multiple languages, resulting in a form of "imposed multilingualism". For example, software and global web companies are under pressure to release their contents in multiple language editions. Others have questioned whether linguistic localization of the web can really minimize language barriers (Lenihan, 2011; Thurlow, 2012; see also Chapter 6). Dor (2004) predicted that "the Net is going to be a predominantly non-English-language medium" (99), which encourages the use of languages other than English online. Global Reach (2003, cited in Paolillo, 2007: 421) further predicted that Chinese would overtake English as the largest language of the internet by 2015. Apparently, this has not been realized yet. Despite advanced technologies such as Chinese inputting, domain names in Chinese, and machine translation, Chinese languages will not replace English in the foreseeable future for political and societal reasons. Politically, the internet is still governed by the US government, who makes ultimate decisions on domain name assignment. Domain names in non-Roman languages are formed much slower than those with Roman scripts, which hinders the development of "an internet culture" in countries that do not adopt Roman scripts. Socially, English is still the official language in many domains, such as scientific publications, business, and universities. It is also increasingly the official second language of many countries (Flammia and Saunders, 2007). For these reasons, it is not surprising that many web users still select English as a more neutral language on the internet (Kelly-Holmes, 2014).

For linguists, the massive growth of internet contents in multiple languages means that new data are available for investigations of computer-mediated discourse beyond the English-speaking context. A growing body of academic research has emerged to provide illustrative examples for multilingualism on the internet. The studies reported in *The Multilingual Internet* a decade ago, for example, were published with the premise that over two-thirds of internet users did not speak English as their first language (Danet and Herring, 2007). Earlier research on multilingualism in CMC tended to take a largely descriptive approach, with an aim of documenting multilingual participants' creative use of orthographic or typographic features in CMC in light of "Netspeak" features identified in English data (e.g. Lee, 2002; Nishimura, 2003; Palfreyman and Al-Khalil, 2003; Tseliga, 2007). For example, Nishimura (2003) shows how Japanese BBS users and their English counterparts adopted a similar set of Netspeak features, such as multiple punctuation, eccentric spelling, capital letters, and so on. In a study of linguistic practices of email and ICQ instant messages by college students in Hong Kong, I categorized the English short forms in the data using Crystal's (2001) classification system (Lee (2007b)). A characteristic that is frequently reported in these descriptive studies of multilingual CMC is the "spelling" or Romanization of words in languages that are not typically written in the Roman script, such as Arabic (Warschauer et al., 2007), Cantonese (Lee, 2007b), Greek (Tseliga, 2007), and Thai (Seargeant and Tagg, 2011). Other multilingual CMC researchers have taken up established concepts in multilingualism in spoken discourse, such as code-switching, which is further discussed in Chapter 3.

While there may be reasons to believe that English is still the largest language on the internet in terms of users and contents, Paolillo (2007) acknowledges that estimates of internet user populations are far from sufficient in representing multilingualism online, and he calls for more empirical research on actual instances of language use online. Similarly, a study by Kelly-Holmes (2004) reveals the complexity of understanding multilingualism on the internet. She surveyed the linguistic repertoires of over 2,000 English-educated bilingual young people in eight different countries, including their competence in English and their national languages. The respondents were also asked to report on their internet language use in different domains, including information search for academic purpose, emailing, and chatting. Although no significant shift to English was noted in the group as a whole, the study shows that speakers of languages that have smaller number of speakers, such as Indonesian and Macedonian, are more likely to shift to English on the internet. By contrast, speakers of languages that are considered to be prestigious (e.g. language of schooling) and have a larger number of speakers, such as French and Italian, are more likely to retain these languages in their online activities. This again challenges the popular belief that the internet is a space where English prevails. However, Kelly-Holmes also warns that the internet "could marginalize some lesser-used languages even further" (2004: 74). This is echoed by the conclusions made by Paolillo (2007) and Deumert (2014) that although English may no longer dominate, the multilingual internet is still very much shaped by a few larger languages.

Understanding multilingualism and its related topics, such as linguistic inequality, is certainly not a straightforward matter; multilingualism online cannot be discussed independently from other social issues, such as language policy and access to information. In 2011, the United Nations declared that having access to the internet is a basic human right (La Rue, 2011), when in fact, by the end of 2014, less than half of the world's population were online (Internet World Stats, 2015). On the other hand, while quantitative surveys may suggest that smaller languages have been disadvantaged, the languages concerned may as well be underrepresented due to the limitations of the survey methodologies. For one thing, these figures are only indicative of what happens to *public* web contents (i.e. whatever content that is on display on the web), and they are heavily reliant on data provided by participants' self-reported information in questionnaires. Data elicited from direct contact with web users is missing from most market surveys and reports. A more comprehensive investigation of linguistic diversity online should involve identifying and understanding the range of meaning-making resources that web users deploy in authentic contexts of communication. This can be achieved through collecting and analyzing logs of CMC exchanges from different platforms. The rest of this chapter thus focuses on people's *language choice* or linguistic preference online by adopting the concept of *resources*. The discussion also draws on samples of authentic digital writing from studies by myself and others on different CMC platforms.

LANGUAGE CHOICE AND MULTILINGUAL RESOURCES ONLINE

Language choice is mostly concerned with the codes or linguistic *resources* available to online participants and how they negotiate their code preferences when communicating with others who may or may not share these resources, regardless of their competence in the resources concerned. In this chapter, language choice is deliberately distinguished from code-switching (CS). CS is more concerned with the process and representation strategy of an interaction. Because CS is found to be a salient resource which serves a wide range of discourse functions in multilingual online interaction (Androutsopoulos, 2013a), it deserves a discussion in its own right and will be the focus of Chapter 3.

Studies of language choice on the internet may begin with identifying the languages or resources available to users. *Resources*, as defined in Chapter 1, consist of aspects of knowledge available to language users in meaning-making. *Available resources* do not exist only inherently and biologically in human beings, nor are they predetermined as something that people "should" know. Rather, the amount of resources available is socially constructed; it relies on a number of variables or factors, such as experience, beliefs, values, and social backgrounds of the person in question (Ivanić and Tseng, 2005). Figure 2.1 is an example of the possible range of meaning-making resources drawn on by a multilingual internet user.

Representational resources include languages, scripts, and various modes through which a message is produced. Resources of *languages* are not just about human language as a mode of communication in general. This chapter is particularly interested in the various identifiable languages and systems of representation that are available to multilingual web users. These may include a national language in spoken or written form, a language learned and used mainly in school, and so on. The available *scripts* or *writing systems* include alphabetic and character writing

Representational resources

- Languages (including regional and social varieties)
- Genres
- Scripts (e.g. Romanization, characters, self-generated scripts)
- Modes, such as
  - Writing
  - Speech
  - Visual images
  - Animations
  - Voice
  - Gestures

Human resources

- Interactants (e.g. recipients of messages)
- Mediators
- Other people who contribute directly or indirectly to the text-making of IM

Ideological resources

- People's perceptions, including values, thoughts, feelings, and beliefs
- People's everyday experiences in using texts

Technological resources

- Software (e.g. chat applications, inputting systems)
- Hardware (e.g. computer, smartphone, tablet, etc.)
- Access to the internet
- Network resources (copy/paste language, online translators)

Figure 2.1 Potential resources for multilingual text-making online

systems, and also some invented systems of writing which people come across in non-CMC contexts. This breakdown of linguistic resources is particularly useful in studying text-making practices in multilingual contexts such as Hong Kong. For example, in my earlier study of email and instant messaging (Lee, 2007b), I identified five major linguistic resources available to the Hong Kong participants: (1) (attempted) Standard English, (2) standard written Chinese, (3) Cantonese in characters, (4) Romanized Cantonese, and (5) morpheme-by-morpheme literal translation. The first type is called *attempted Standard English* because it is clear from the immediate context that the sender intends to write in Standard English (such as the variety taught in schools), but the resulting message may contain characteristics of spelling and grammar that are not commonly expected in standard written language. In Example 2.1, the sender, a student assistant to Dr. Leung, attempts to write in rather formal and Standard English. The email, however, is infiltrated with features that are not typically associated with standardness, such as the short form *u* for *you*.

Example 2.1 An email written in attempted Standard English

Dear Dr. Leung,

I would like to let u know that I will continue my work on Friday morning. I have already transfer all the files to PDF formate. . . .

The second type is *standard written Chinese* represented with standard traditional characters. This is also the form of written Chinese that is taught in school and used in most official and formal domains, including the government. In addition to standard forms, the study found that the Hong Kong participants have created innovative ways of representing Cantonese in their IM chat (types 3, 4, and 5). While it is possible to word-process Cantonese characters, extra technical support and effort are still required, such as installing a supplementary character set. For convenience, some participants borrow homophonous characters from standard written Chinese to replace a difficult or unknown Cantonese character, that is, using characters that are different in form but have the same pronunciation as the intended Cantonese characters, such as using 野 (*je5*) to represent 嘢 (*je3*). Here, the two pronunciations differ only in tone. A similar strategy is also noted in Su's (2007) research on "stylized" Taiwanese on a Taiwan-based bulletin board system. Another common way of representing Cantonese without using Cantonese characters is to spell out Cantonese words as Romanized words, notably sentence final particles (as also noted in James, 2001) or idiomatic expressions. Often, when a longer Romanized Cantonese expression is used, quotation marks are inserted to distinguish the expression from the Standard English in the rest of the message, thus indicating that the quoted expression is a "marked" choice. For example, in *then "bei sum gei" la cu next time*, the expression *bei sum gei* is the writer's self-invented spelling of 比心機, which literally means "hang in there". In the study, the most commonly adopted representation resource for Cantonese is English *transliteration*, that is, the direct translation of Cantonese into English in a morpheme-by-morpheme manner, and the resulting expression often makes no sense to native English speakers. An interesting example of transliteration is the phrase "sky and land lessons", a common phrase among university students to mean a long period of spare time between an early lecture in the morning and another one in late afternoon. It is a literal translation of the Cantonese expression 天 (sky) 地 (land) 堂 (lesson). Such creative and strategic mingling or "meshing" of Chinese and English scripts to express Cantonese online certainly gives rise to new multilingual practices and encounters online.

By providing a finer-grained classification of linguistic resources, I have also moved beyond the conventional meaning of *codes* from referring to broad categories of languages (such as Chinese, English, French) to including different representational resources for these languages. Such a refinement is useful and necessary, because it is often in the production of online texts where these non-standard yet creative forms of meaning-making resources are drawn on simultaneously. Other kinds of resources also contribute to people's multilingual practices online. In online contexts, as listed in Figure 2.1, written language is often complemented by other modes of representation, such as images and videos. A typical example of visual resources in CMC would be emoticons, which may be typed out or inserted from preinstalled applications. They may be used independently or combined with words to form multimodal messages (see Derks et al., 2007; Dresner and Herring, 2010; Deumert, 2014), as discussed further in the next section.

In addition to modes of representation, *human resources*, such as other online participants and mediators (e.g. someone who helps translate contents into English), are crucial in various kinds of decision making involved in multilingual text-making practices. *Ideological resources* refer to perceptions, thoughts, feelings, beliefs, and other values associated with language choice and other resources. Of course, various *technological resources*, such as software and hardware, are essential for all kinds of digital communication. Androutsopoulos (2013a) also draws attention to the increasing fluidity of digital resources over networked spaces, especially social media. Online translation services such as Google Translate and Facebook's "see translation" function allow web users to produce texts in a language of which they have no knowledge. Flows of network resources also give rise to what Androutsopoulos refers to as "copy/paste language", such as copying song lyrics from YouTube and pasting them onto one's Facebook wall with additional annotations in one's local language. These network resources together bring the dynamics of multilingual practices to the forefront.

Amongst these resources, some are more open to reshaping and transformation, such as languages and scripts; others are drawn upon consistently across contexts, such as technological resources. Despite the various multimedia possibilities offered by digital media, interaction online relies heavily on the written word (Merchant, 2006). Languages, scripts, and other representational modes are considered to be crucial resources in online text-making. It is important to note that Figure 2.1 is not intended to be an exhaustive set of "repertoires" for all multilingual internet users, nor is it meant to remain static over time. New resources may be introduced to one's linguistic ecology and existing ones may be reshaped, reappropriated, or recontextualized depending on the contexts of use.

Deploying linguistic resources

As English has been foregrounded in the literature of online multilingualism discussed at the beginning of the chapter, it is worth zooming in to English as a linguistic resource in multilingual activities on the internet. The perceived dominance of English is best demonstrated by global web users' preference for English as *the* lingua franca on the internet. Language shift to English is not uncommon among online users who do not share the same first language. In a Swiss medical students' email list, "Pan-Swiss English", a form of nonstandard English heavily influenced by Swiss languages, is found to be the dominant language of interaction among a group of French- and German-speaking students (Durham, 2007). Example 2.2 is an email written in Pan-Swiss English from Durham (2007). Words or expressions underlined are considered by Durham to be non-native English features.

Example 2.2 An email written in Pan-Swiss English

Hi guys, you remember at the past meeting in Bern a red map that 'b' gave to 'c', concerning the activities of SCOPH and the presentation of their activity? Well, 'c' (man! such a stupid NEO!!!) has lost it somewhere. Has anybody of the participants of the meeting taken it home? Please, answer quickly! I need that map desperately!

A lot of THANKS

According to Durham's study participants, English was a natural choice due to its commonly accepted lingua franca status in Switzerland. Durham (2007) adds that it is indeed "impractical and confusing" to mix languages in a public mailing list for which the primary aim is to communicate with a wider audience that does not share the same home language.

The preference for English as a lingua franca is further reinforced on globalized media such as the photo-sharing site Flickr. What follows is a more detailed discussion of findings about the role of English from a study of multilingualism on Flickr. Lee and Barton (2011) analyzed 100 Flickr sites for their multilingual contents as well as users' online text-making practices. Of the 100 sites observed, 51% of them contained English only contents (including captions, tags, comments), whereas over 70% of the user profiles were in English only. A further analysis of 18 Chinese users' profiles revealed that all of the profiles are written either in English only or in a mixture of Chinese and English. For example, "JadeCastle", a professional photographer based in Taiwan who speaks primarily Mandarin Chinese, welcomes his visitors in English on his profile page (Example 2.3).

Example 2.3 An English profile by a Taiwanese user on Flickr

About JadeCastle

I live in Taipei and enjoy taking pictures as my favorite hobby.
Welcome to visit my photo blog at: jadecaso.blog.com/

Elsewhere on his site, JadeCastle also writes captions and tags in English only, for example, "sunrise", "tide", "end of summer". In an email interview, when asked why he used English extensively on his site, he explicitly referred to the global status of the language: "I believe English is better for 'international' audiences [on Flickr]". As a professional photographer, JadeCastle is keen to connect with international photographers and share his interest in photography through his Flickr site. Similar ideas about the global or universal status of English were frequently articulated in the interviews with other Chinese participants, who might come from Hong Kong (where most people speak Cantonese and write in traditional Chinese), mainland China (where many people speak Mandarin and write in simplified Chinese and Romanized Chinese/hanyu pinyin), and Taiwan (where most people speak Mandarin and write in traditional Chinese). This also implies that the decision of using English on Flickr may be unrelated to users' competence in the language. As remarked by many of the Chinese participants, they would rarely use English outside Flickr, despite the fact that there was a strong presence of English on their Flickr sites.

When several languages coexist on a webpage, where they appear on the page may be indicative of the different degrees of significance given to the languages by the site designer (Kress and van Leeuwen, 1996). The emphasis on English is evident in the placement of English and Chinese contents in the profile of "tiong": "My English is poor, so I take photos". Underneath this is a line in Chinese: "唉,中文都一樣. . . . 不見得是好" (*Translation: well, . . . so is my Chinese . . . it isn't good either*). To tiong, each of the two languages carries specific social and cultural meanings. Tiong is a Hong Kong Chinese who claims to use more Chinese than English in his everyday life. In this profile message, although acknowledging that he does not master Chinese or English, the fact that the English text is placed above the Chinese writing fits in well with what Scollon and Scollon (2003) refer to as "indexicality" of code preference. That is, the fact that he places English on top reveals that English is what Kress and van Leeuwen (1996) call "ideal" to tiong, and it is the language that he expects the world to see first on his profile page. This also illustrates his belief that English is the common language shared by his international audiences on Flickr. Similar patterns were also identified among Spanish–English bilinguals (see also Barton and Lee, 2013).

For the cases of the Swiss mailing list and Flickr, English primarily serves as a lingua franca for mutual comprehensibility. Sometimes, people's preference for English in CMC may be unrelated to achieving mutual understanding because it has been used among participants who *do* share the same native language. In Egypt, English is the main language in both formal and informal forms of CMC among Egyptian internet users. Warschauer et al. (2007) note that the Egyptian professionals in their study tend to use English only in email, especially in formal communication. By contrast, the common form of writing in Egypt, classical Arabic in Arabic script, is almost absent in all of their data sets. Egyptian Arabic, which has no developed writing system, is found to be quite prominent in "highly personal content that they could not express well in English" (Warschauer et al., 2007: 312). The participants do not view their preference for English as abandoning Arabic culture; rather, they take pride of their use of English in CMC because the coexistence of various language forms displays Egypt's multicultural history. Once again, knowledge or competence in languages does not immediately shape language choice in online communication. With Warschauer et al.'s study, the mode of communication and formality are important factors behind people's preferences for linguistic resources online.

Language shift to English in CMC among people with a common language is also a reflection of the social status and perceived value of English in a certain community. In a study of SMS texting among bilingual English–isiXhosa users in South Africa, Deumert and Masinyana (2008) report that the majority of their study participants prefer English to their local language in SMS texting. They explain this shift to English in light of the growing social status of English in South Africa because it is playing an increasingly significant role in both educational and domestic domains. Mafu (2004) also notes that in Tanzania, the emphasis on English in the education system is encouraging young people to practice the English they have learned in school. Interestingly, some of my student participants in Hong Kong also remark that they would make use of CMC (such as instant messaging and blogging) to practice their English writing before examinations. With these cases, rather than seeing language shift as a threat to national languages, shifting to English should be understood as CMC users taking up informal and autonomous language learning opportunities offered by the globalized internet (Benson, 2013). Further evidence of multilingual internet users taking up English learning opportunities on the internet can be found on IM (Lam, 2009), fanfiction.net (Black, 2009), and YouTube (Benson and Chan, 2011). I will return to these studies in later chapters.

By contrast, many studies have also reported on cases where dialects and local varieties are maintained and written online. Varieties that are commonly used in spoken contexts are increasingly visible in online communication. Siebenhaar (2006) notes that Swiss-German dialects that rarely appear in offline written communication among the German-speaking Swiss communities are increasingly prominent in IRC chatrooms. On diasporic forums, the "home" language of the website is deliberately maintained by geographically dispersed migrant groups. For example, German is the main language among the Persian, Indian, and Greek migrants on German-based diasporic discussion forums (Androutsopoulos, 2007). In a more recent study, Androutsopoulos (2013c) looks into the presence of German dialects in YouTube videos. Although YouTube is commonly viewed as a global platform, it gives rise to local activities. First, Androutsopoulos observes that many YouTube videos are tagged with German dialect-related keywords such as *Bairisch* (Bavarian), *Alemannisch* (Alemannic), or *Berlinerisch* (Berlin city dialect). Second, these dialects are not only the main languages used to present the videos, but they are also discussed as a theme in these videos and in the viewers' comments underneath. As a result, the increasing visibility of smaller languages and dialects in newer social media certainly fosters more translocal interactions across the globe. The internet has presented both challenges and opportunities for minority languages such as Luxembourgish and Welsh, which will be discussed in Chapter 6.

People's preferences for their national or local languages are also closely related to issues of identities and authentication. Lee and Chau (2015) looked at the functions of Instagram hashtags during the Hong Kong "umbrella movement", a social movement toward universal suffrage, in late 2014. Among the 700 posts collected in their study, 88% were tagged *bilingually* in Chinese (Cantonese) and English, and about 11% of the photos were tagged in Chinese only. English, which is seen as the lingua franca of another photo site, Flickr, became less significant to the Instagram participants in Lee and Chau's study. As few as five posts out of the 700 were annotated with English-only hashtags. This, they argue, is likely to be the result of the increasing tension between the mainland Chinese government and Hong Kong and the pressing need to assert their identity as a "Hongkonger", who is bilingual in spoken Cantonese and English, rather than a "Chinese only" identity. This case will be discussed in greater detail in Chapter 4.

WHEN LINGUISTIC RESOURCES MEET OTHER SEMIOTIC MODES IN CMC

So far, the chapter has focused largely on linguistic resources and the ways in which they are deployed on different platforms based on findings from case studies of different parts of the world. Some of the data in previous research have already suggested that multilingual web users are likely to creatively combine and "stylize" resources of writing (e.g. scripts) and speech (e.g. spoken Cantonese or Egyptian Arabic) in their online writing. However, as shown in Figure 2.1, linguistic resources are only a small part of the rich array of meaning-making resources available to web users. Although the written word plays a vital role in most forms of digital communication, nonverbal modes of communication, especially the visual, have proved to be as significant in meaning-making as written communication. For example, a wide range of pragmatic functions of emoticons are explored in Dresner and Herring (2010), while Sindoni (2014) studies video-chat from a multimodal perspective. In fact, from the moment of logging on to any digital platform, participants inevitably engage in a landscape of semiotic resources. Take the common mobile instant messenger WhatsApp as an example. Before the users can actually send a message, they have to interact with the interface system, which includes a contact list ordered chronologically by time of last contact. Each person may have a profile photo that is optional and can be updated from time to time. The dialogue bubbles on the chat screen are designed in a way that they resemble turn-taking in a conversation. When composing a message, users can not only choose from the keyboard of their self-selected languages, but they are also able to play with a range of emoticons, or emoji icons, of human faces, animals, flags, seasons, food, and so on.

One of the most-cited reasons why visual symbols and emoticons are used in CMC is the lack of visual cues such as facial expressions and gestures in digital writing. At the most basic level, punctuation symbols on the standard QWERTY keyboard can become playful resources for visualizing gestures, as in the formation of basic sideways emoticons such as :-) and >:-<. It has also been noted that some users would visualize their gesture of pointing by adding the arrows (>>) or the equal sign (=) to refer to what follows it. The equal sign is used extensively in my

IM data, such as the following message by a Hong Kong college student Minnie, who was complaining about Aaron's lack of organization skills (Example 2.4).

Example 2.4 An MSN message with the equal sign

Minnie: 交比aaron = 交比rubbish bin
(Translation: Sending [it] to Aaron means sending [it] to a rubbish bin.)

The intended function of the equal sign here may be comparable to that of abbreviations in English, that is, to minimize typing effort by typing one single key instead of the word *means* (requiring the typing of five keys) or even longer expressions. This assemblage of Cantonese writing, English, and an equal sign allows Minnie to effectively soften this potentially offensive message in a playful and succinct manner.

Beyond the standard keyboard symbols, other images have been reported to be prevalent in CMC in Asia. For example, the star signs (☆★) and musical notes (♪) are popular amongst Japanese CMC users (Katsuno and Yano, 2007; Nishimura, 2007). Example 2.5 is taken from a Japanese BBS website, cited in Nishimura (2007: 172).

Example 2.5 A Japanese BBS post with a kaomoji

復活おめでと~♪良かつたね(*∧▽∧*)
(Translation: Congratulations on your comeback [as if singing] that was good [∧▽∧*])*

The emoticon (*∧▽∧*), representing "the mouth wide open, laughing loudly and cheerfully, with asterisks used to indicated rosy cheeks" (Nishimura, 2007: 172), is an example of what is called a *kaomoji* (face mark) in Japanese (Katsuno and Yano 2007; Markman and Oshima, 2007). These face marks are meant to be read upright instead of sideways as in the well-known smiley face :-). According to Nishimura, *kaomojis* are so frequently used in Japan that they have been "codified" in dictionaries. Similar sets of vertical emoticons are also preinstalled in some popular mobile IM apps in Asia, such as Line and WeChat. Culture-specific emoticons have been reported in various studies of Asian uses of emoticons. For example, earlier multilingual CMC research done by Japanese scholars tended to compare Japanese and English uses of emoticons (Markman and Oshima, 2007; Nishimura, 2007). The kaomoji (*∧▽∧*) in the above example is created with Unicode symbols, while other kaomojis can be produced by combining punctuation symbols and *kanji* or Chinese characters. For example, (^^)_旦~~ is believed to resemble someone enjoying a cup of tea (as represented by the character 旦). On Chinese-based CMC, posture emoticons have been invented to express gestures or body postures. For example, orz and 囧rz are said to resemble someone on their hands and knees and looking disappointed (Yang et al., 2007). In orz, o represents the head, r is the arms and the torso, and z is the legs. The 囧 in 囧rz is believed to look like a rather distressed or disappointed facial expression (Wikipedia, 2015), as illustrated in Example 2.6, which is a post by a user in a Taiwan-based BBS (Chang, 2009: 39).

Example 2.6 A Taiwanese BBS post

Could someone tell me do zoloft and rivotril have strong side effects? I lose all appetite and feel like throwing up sometimes. I am really depressed. . . . 囧rz

Both orz and 囧rz are becoming more visible on globalized websites such as Twitter and other social media as they are gaining popularity among Asian web users. (See a Twitter feed featuring the hashtag 囧rz: https://twitter.com/hashtag/%E5%9B%A7rz). In addition, tweets tagged with these posture emoticons are more likely to be written in Asian languages including Chinese, Japanese, and Thai.

To a certain extent, the case of Asian emoticons is comparable to the creative deployment of scripts discussed earlier. First, the creation and use of kaomoji and posture emoticons reveal playful manipulation of resources. Playfulness in CMC helps create a sense of intimacy and familiarity between users (Katsuno and Yano, 2007). Second, like the resources of languages and scripts, visual resources such as emoticons are given meanings at the time of chatting and in the context of different groups of users, where interpretations of meanings also take place. When emoticons are taken away from a particular context, their meanings might vary.

From the users' perspective, the meanings of emoticons are not static. In my previous research on IM, I often found myself checking with my participants on the meanings of the emoticons used in their chat logs, and surprisingly, on several occasions, we could not reach a consensus of what certain emoticons meant. As another example, a friend of mine has once complained about my "inappropriate" use of emojis in my WhatsApp messages, when in fact she has misinterpreted my use of the "tears of joy" emoji as a crying face. Even with the most basic smiling face emoji, its meaning must be interpreted within the immediate situation of use. In an exploratory study of the meanings of emojis in the Swedish context, Kelly (2015) asked 90 student respondents to give their interpretations of the message: "I miss you [smiling emoji]". Interestingly, the answers ranged from seeing the writer as being "sarcastic" (19%) to being "honest and sincere" (26%).

The fact that emoticons or emojis have been understood in different ways across cultures and individuals reveals that the meanings of emoticons are not universal, but situated within users' practices. Although attempts have been made to provide taxonomies of the pragmatic functions of emoticons (e.g. Yang et al., 2007; Yus, 2014), and some have even gone so far as to argue for a universal visual language of emoticons (Azuma and Ebner, 2007), there is in fact no fixed set of "conventional" meanings and uses of these symbols. This is also true of all other representational resources in digital writing, including languages and their associated scripts. As Deumert (2014) argues, the linguistic or so-called Netspeak features that have been taken as norms in English-based CMC, such as abbreviations and phonetic respellings, are "realized differently in different places. They are not simply a replication of a global (English) norm, but become local practice" (Deumert, 2014: 141). Digital writing, she further argues, "is not about *rules*, but about strategies and the display of creativity" (142). This leads to the need of moving beyond generating frequency patterns to further explore the ways in which the use of resources is situated within the social practice of digital communication.

UNDERSTANDING THE AFFORDANCES OF MEANING-MAKING RESOURCES

In any study of multilingualism on the internet, identifying the linguistic resources drawn upon can be a useful point of departure for researchers to gain a snapshot overview of the data at hand. As noted by Androutsopoulos (2007: 359), the mapping of language choice in CMC "prepares the ground for analysis of particular online interactions". However, staying at the level of identifying languages and code choices can be limiting. An analysis solely informed by a language choice approach is especially problematic when it comes to interaction involving multimodal resources, as demonstrated in Example 2.7, which is an IM conversation between two Hong Kong university students.

***Example 2.7* An MSN conversation in emoticons**

1 Shine:	when are you free ar		
2 ABC:		[smiling]	
3 Shine:		[sarcastic]	
4 ABC:		[hugging]	
5 Shine:	i treat you ar		
6 ABC:	thx, bro . . . but, I am busy late ly ar		
7 Shine:	. . . .	[silence]	
8 ABC:		[hugs]	
9 Shine:		[crying]	
10 ABC:		[hugs]	
11 Shine:		[broken heart]	
12 ABC:		[crying]	
13 Shine:		[doubting]	
14 ABC:		[hugs]	
15 Shine:		[dying rose]	
16 ABC:		[crying]	
17 Shine:	can you use mp3 r/t gar		*(Translation: Can you use mp3 ringtones?)*

This IM exchange is predominantly emoticon-based. ABC does not immediately respond to Shine's question in turn 1. Instead, he replies with just a single emoticon without any words (or what Provine et al., 2007, call "naked emoticon"). And in return, Shine replies with another emoticon. Toward the end of the conversation (i.e. turns 8–16) there is a continuous exchange of naked emoticons until Shine initiates a new topic in turn 17. Earlier approaches to multilingual CMC would focus on the words (turns 1, 5, 6, and 17) and describe this exchange in terms of a new variety of e-discourse (Fung and Carter, 2007) or even a form of Hong Kong English (Bolton, 2000) with features of Cantonese spelling inserted into an English utterance. A language choice approach would first identify the codes available in this exchange. For example, each of lines 1, 5, 6, and 17 consists of an English-based utterance ending with a Romanized Cantonese discourse particle (ar), while the rest of the interaction is characterized by a series of emoticons. These generalizations, however, can be problematic because they would leave a range of questions unanswered: Why would both ABC and Shine rely largely on emoticons in this exchange? Why must the writers combine Cantonese spellings with English instead of writing in just Standard English? Do the writers draw on the same range of resources in all contexts of IM exchanges and beyond? Apparently, the particular styles of writing serve specific and situated social meanings and index specific kinds of writers and recipients. What is most intriguing in this example is not just the wide-ranging forms of text-making resources, but how and why these resources are deployed for different purposes by different writers. It was only through interviewing ABC that the reason behind their exchanges of emoticons became clear:

> ABC: actually there wasn't much to say. So when he [Shine] sent me an emoticon, I replied with an emoticon. Sometimes it's just to let my friends know that I'm online. It could be an "angry" emoticon but that doesn't mean he/she is really angry. They just want to let me know that they're online.

In reality, not all of these forms of language and resources are preferred in any given situation. Sometimes people would use one resource only, while at other times they would mix linguistic codes. There is no fixed pattern in representing, using, or choosing the resources. This is a result of the participants taking up and acting upon the *perceived affordances* (Lee, 2007c), or the action possibilities and constraints, of the various resources available to them.

In working out what different languages can or cannot do in CMC, online participants take into consideration a number of ecological factors. These factors cannot be arrived at by observing chat logs or digital texts alone, but are a result of getting closer to web users through qualitative research such as interviews and participant observation. Some of the more salient factors include the following:

- *Situated language ecology of individual users:* This refers to online participants' sociolinguistic and cultural backgrounds, their everyday experiences, and attitudes and perceptions toward different linguistic resources, such as different scripts. Understanding people's technolinguistic biographies can be a useful starting point. From my study of IM through to my recent study of social media or Web 2.0 practices in Hong Kong, many of the student participants have associated both Standard English and standard Chinese with seriousness or formal writing. The students said they would reserve these standard forms of written language for emails with professors or when they post about more serious feelings on Facebook or their blogs.
- *Target audience:* The intended reader of a message or any digital writing may shape one's language choice. In digital communication, one's target audience may include the general unknown public viewers (e.g. commenters on

YouTube or blogs) or online audience of people whom one may or may not know in real life (e.g. friends and friends of friends on Facebook). Participants in the Flickr study, for example, often pointed out that describing or tagging their photos in English was to reach a wider and global audience (Barton, 2015). Young people in Finland have reported a similar factor about their use of English in fan fiction writing (Leppänen et al., 2009).

- *Content of post*: What is posted can directly affect the language choice of both the writer and the audience. On Wikipedia, topics which are believed to be of local significance tend to be written in a particular language only and may not be available in English. For example, the colloquial Hong Kong Cantonese expression 暗串 (being implicitly sarcastic) is available only in the Cantonese edition. On Instagram, on the other hand, images about the Hong Kong umbrella movement are more likely to have bilingual hashtags (Lee and Chau, 2015; see Chapter 4).
- *Technological possibilities and constraints*: These refer to the various functions available on a certain CMC platform, one's familiarity with different inputting methods, convenience of software and hardware, and so forth. For Greek web users, it is not uncommon to use "Greeklish", a Romanized form of Greek script, because encoding Greek characters is more time-consuming than typing in Greeklish (Spilioti, 2009). A similar trend also exists among Cantonese–English bilingual web users (Lee, 2007c).

All these factors are interrelated and may be taken into account one at a time or in combination. These and other related factors are discussed in greater detail in Lee (2007c). A similar set of factors is also applied to explain code choice between English and other languages in newer social media such as Flickr (Barton and Lee, 2013).

BEYOND COUNTING LANGUAGES

The claim that English is the largest language on the web does not accurately represent what is happening to languages on the web. Global statistics and quantitative surveys tend to suggest that English still plays a significant role in online communication, but it no longer dominates the internet. Other languages, including smaller languages, are represented on the web to different extents and in different ways. Whether people's strong preference for English would automatically lead to linguistic inequality online is a complex matter; as already mentioned, this is related to a set of factors including online and offline language policies, level of access to information, and resources available to individual users. In any given digital communication context, some factors become more salient than the others.

From the point of view of web users, English is not necessarily in competition with other languages. As we have seen, English may be the lingua franca to some users in one episode of online interaction, while at other times a local language is preferred. The cases already reveal huge variation of language choice across CMC modes, language contexts, and user groups. Studies have also shown that to those users who do have regular access to the internet, writing in a language that is not their national language on the internet is not necessarily a negative experience. They are indeed comfortable in taking up the agency and playing with the affordances of the multiple linguistic resources available to them. In studies of language choice that adopt an interview or user-oriented research method, almost no web users have expressed anything negative about their shift to English. By contrast, the web has opened up a space for them to explore and play with their linguistic and multimodal resources creatively and strategically, something they would rarely do in other contexts of writing. Therefore, rather than debating whether English still dominates the internet or whether "internet English" exists as an independent variety, what is more important and meaningful is to examine how English and other languages are negotiated and deployed strategically in a world of dynamic digital media.

This chapter has focused on language choice where linguistic resources are discussed separately in a given study. In reality, these codes are not used one at a time but coexist in the same discourse. The next chapter looks at another common area of multilingual CMC: code-switching.

3

WRITTEN CODE-SWITCHING ONLINE

Overview

- Defining codes and code-switching in digital communication
- Research on code-switching in digital communication
- Structural patterns of code-switching online
- Discourse functions and social motivations of online code-switching
- Rethinking code choice and code-switching in the digital age

Code-switching (CS) is a common linguistic phenomenon that is observed in many multilingual contexts. In multilingual CMC research, the theme of code-switching is often pursued alongside language choice. Examples 3.1 and 3.2 are taken from two different studies of CS in computer-mediated discourse. Example 3.1 is a post from a thread on theinder.net, a discussion forum dedicated to German-based migrants from India studied by Androutsopoulos (2007, 2013c). The post is written predominantly in German, ending with a question asked in English. Example 3.2 is a "weather message" on a Chinese government-run microblog called Shanghai Release (Zhang, 2015). Different from the German post, this example exhibits a more even deployment of Chinese, represented in simplified characters, and English.

Example 3.1 A post on theinder.net (German and English)

im westen ist es auch tradition jungfräulich in die ehe zu treten!!! das tun auch einige (bsp. spanierinnen, italienerinnen, etc..)! wieso wird immer der westen für alles verantwortlich gemacht??? is there no gravity in indian brains?

(Translation: In the west, no sex before marriage is also a tradition!!! And some stick to it as well [e.g. Spaniards, Italians, etc.]! Why is the West always being blamed for everything??? Is there no gravity in Indian brains?')

(Androutsopoulos, 2013b: 681)

Example 3.2 A post on Shanghai Release (simplified Chinese, English)

多云中午前后转阴有雨,局部大雨,26–32度: #早安上海# Everybody, it's holiday today. Are you ready? today 又是 Sunday, 我好想 day day 都非 hot day,Where good where cool 我们 where stay, NO 冷 NO热, Love 谁谁. It will be rainy! I'm sorry. 冷空气南下与暖湿气流交汇,雨水难逃好乐 day

(Translation: Cloudy to rain around noon, isolated heavy rain, 26–32 degrees: #Good morning Shanghai# Everybody, it's holiday today. Are you ready? Today is Sunday again, I do not want to have hot day everyday; we stay at wherever is good and cool, NO cold NO hot, Love who [who cares]. It will be rainy! I'm sorry. Cold air will land in the south and meet warm and moist airflow, hard to avoid rainfall on holiday.)

(Zhang, 2015: 243)

The first thing that these two posts illustrate, obviously, is that multilingual internet users do not always use one language at a time, nor do they use one language in one online message. Second, as with spoken CS, there exist different structural patterns and styles of CS in CMC. Some styles may be exclusive to CMC contexts and may not be accounted for by existing frameworks of CS in speech. Third, the two examples of CS serve rather different discourse functions. Switching to English in the premarital sex discussion in Example 3.1 allows the writer to emphasize the conclusion made; whereas in the Shanghai Release example, alternating between Chinese and English allows the writer to express playfulness and offer practical information about the weather at the same time. The wordplay in Example 3.2 is in fact a playful adaptation of the lyrics of a viral mixed-language pop song called 好樂 day ("Holiday"). The examples to be discussed in this chapter mainly cover these three areas of CS in CMC: (1) the co-existence of different linguistic codes in the same CMC message, (2) different structural patterns of CS, and (3) different discourse functions of CS. The rest of the chapter first offers an overview of the meanings of CS within the context of computer-mediated discourse. It then moves on to provide a survey of existing research in the literature of CS in CMC. Using examples from various studies and platforms, the chapter also identifies common styles and discourse functions of CS in computer-mediated discourse.

DEFINING CODES AND CODE-SWITCHING IN DIGITAL COMMUNICATION

In sociolinguistics, a *code* is commonly understood as a language or a variety of a language. In the context of this book, a code is seen as a form of linguistic resource (see Chapter 2) that participants draw upon for meaning-making in any situation. Traditionally, CS is defined in terms of structural features and conversational patterns. One of the most cited authors of spoken CS is John Gumperz, who defines CS as the "juxtaposition within the same speech exchange of passages of speech belonging to two different grammatical systems or subsystems" (1982: 59). Gumperz also introduced the concepts of *we-code* and *they-code* in the context of bilingual communities involving minority languages. According to him, we-code refers to a minority language used in informal situations to assert in-group identities, while they-code is reserved for formal contexts. We- and they-codes have been applied extensively in the literature to other bilingual communicative contexts.

Later conceptualizations of CS tended to build upon Gumperz's ideas. Carol Myers-Scotton's (1993) Matrix Language Frame model is another popular framework of CS in speech. The model posits that in CS, one language serves as the matrix, or base language, to form the grammatical basis for the mixed-language sentence. For example, in Hong Kong, Cantonese tends to be used as the base language to which English is inserted in code-mixed spoken utterances produced by Hongkongers (Li, 2000). This is not always the case with computer-mediated discourse, as illustrated in some of the examples in this chapter. Myers-Scotton's markedness model also proposes that code-switching, or language choice more generally, may be an expected (unmarked) or unexpected (marked) choice.

Also commonly applied in CS research is Peter Auer's (1999) conversational approach. In this approach, Auer provides a classification system for conversational CS, including insertional switching and alternational switching. In insertional switching, one language serves as the base language to which words from another language are inserted. Alternational switching involves switching between languages across sentences (see the section "Structural patterns of code-switching online" for details). In the literature, CS is often contrasted with *borrowing* (see Spilioti, 2009; Myers-Scotton, 2010), a process through which a lexical item from another language is assimilated into a language that is in use. This often includes names of people and places, cultural-specific items (e.g. SMS, internet), and formulaic expressions (e.g. hi, sorry). Traditionally, in a code-switching situation, the speaker is expected to be proficient in the languages in use; however, in a borrowing situation, the borrowed words are used not because the speaker is more familiar with or prefers this language, but because those expressions have already been used as a part of the language. In other words, borrowed words are usually more recurrent, whereas code-switched items are not (Myers-Scotton, 1992). For example, the insertion of the English word *so* in Spanish is considered to be borrowing because it is speakers' unconscious choice (Lipski, 2005). The use of the English word *internet* in many languages is also a case of borrowing for the same reason. The distinction between code-switching and borrowing is problematic, as there is no adequate evidence suggesting whether such uses are conscious. For this reason, this book does not insist on such a distinction.

In online communication, code-switching can be broadly defined as a single writer drawing upon multiple (i.e. two or more) linguistic resources in a single discourse or multiple discourses. In discussing the structure of code-switching in computer-mediated discourse, I take a narrow view of discourse to refer to a linguistic unit used by people in social contexts, while acknowledging that digital discourse is also a social practice (Androutsopoulos, 2012). In CMC, CS has more specific meanings than simply the coexistence of multiple languages. Online code-switching is a more restrictive form of multilingual CMC in that the different languages used are "dialogically interrelated by responding to previous, and contextualizing subsequent, contributions" (Androutsopoulos, 2013b: 673). In other words, any instance of CS contributes to the overall interactivity of an online exchange. According to this definition, then, someone using one language for a Flickr photo caption and responding to a viewer's comment in another language may not necessarily be a case of switching, as the two may not be "dialogically interrelated". However, as some of my examples will show, whether something qualifies as a case of code-switching is not always determined solely by its structure.

Theories and generalizations about spoken CS in the work of Gumperz, Myers-Scotton, and Auer have been taken up extensively by CMC researchers, as one of an early assumptions about language use in CMC is its resemblance to spoken discourse (see, for example, Ferrara et al.'s [1991] idea of "interactive written discourse"). To name a few examples, Hinrichs's (2006) work on CS in email draws on Myers-Scotton's markedness model and Gumperz's ideas of we-code and they-code and CS as contexualization cues. Auer's classification of CS patterns is applied in Leppänen's (2007) study of the CS practices among Finnish young people and in Siebenhaar's (2006) study of Swiss-German IRC chatrooms.

These models developed for spoken interaction, however, may not fit in perfectly with CS in written texts such as CMC (Sebba, 2012; Androutsopoulos, 2013a). For example, the markedness model may not work with a playful yet informative weather forecast message as in Example 3.2 because, for one thing, the base language in this post is unclear. Another notable feature of CS in digital writing that cannot be accounted for by a spoken CS model is the alternation between writing systems or scripts. *Script-switching* involves the alternation between orthographic resources. The expression "好乐 day" in Example 3.2 is a case of creative assemblage of Chinese and Roman scripts to represent the English word *holiday*, in which the pronunciation of 好乐 resembles the pronunciation of *holi-* in *holiday*. Other examples of script-switching in CMC include switching between Arabic and English scripts in Palfreyman and Al Khalil's (2003) and Huang's (2009) finding on switching between Chinese and English.

The term *code-switching* (instead of code-mixing) is adopted in this book. On the one hand, it is a cover term for both insertional and alternational switches between codes at all structural levels, including switching across sentences (*intersentential switching*), switching within a sentence (*intrasentential switching*), and switching within a word (*intraword switching*). More importantly, CS is considered to be a more neutral notion, especially in bilingual educational contexts such as Hong Kong, where mixing is often stigmatized and discouraged (Li, 1998). As Li (1999: 7) suggests, "[to] avoid negative connotations associated with the term 'code-mixing'", the more general term code-switching is used in this book to "cover switching at both the inter- and intra-sentential levels.

> . . ."cover switching at both the inter- and intra-sentential levels. There is, however, no other satisfactory term which can replace the substantive '*mixed code*' and the adjective '*code-mixed*' "

Auer (1998) makes a clearer distinction between switching and mixing in that, compared to switching, mixing serves more stylistic functions than pragmatic functions. According to this distinction, the occasional insertion of *baby* in many Cantonese love songs is a stylistic choice rather than an actual need to communicate by switching to English. This chapter does not foreground such a distinction, as there are cases where switching and mixing become unclear (as in the weather forecast in Example 3.2).

RESEARCH ON CODE-SWITCHING IN DIGITAL COMMUNICATION

The CS models discussed in the previous section are not always applicable to written texts and online discourse. A key motivation behind studying CS in online communication is the significant imbalance between studies of spoken CS and written CS (Sebba, 2012; Androutsopoulos, 2013a). Researching CS in writing is not a new idea; there were studies of written representations of speech such as dialogues in literary works and written records of conversations as early as the 1960s and 1970s (Sebba, 2012). The rise of digital written communication has certainly provided a new landscape for the exploration of CS in written language in everyday life. This has also resulted in an increased interest in researching multilingualism in online communication. This section provides an overview of the range of research contexts in CS research in CMC.

Androutsopoulos (2013b: 674) offers a concise summary of a selection of 20 studies of CS in CMC between 1996 and 2009. The summary lists the 20 works by authors, modes, forms of participation (public or private), languages, participants and their social settings, and methodology (quantitative or qualitative). Following the structure of this summary, Table 3.1 is an updated selection of 25 studies of CS in CMC, including those published after 2009. The publications are ordered chronologically by publication year. Only studies that explicitly treat CS as a theme are covered here. The list also includes works that problematize the concept of CS. These studies often have CS as a starting point and then move beyond CS to newer concepts such as code-meshing and polylanguaging (e.g. Jaworska, 2014, on language play on forums and Thorne and Ivković, 2015, on YouTube comments). Most of the researchers who are interested in CS are themselves multilingual or come from diverse linguistic and cultural backgrounds. Many study their native languages or other languages with which they are familiar. Studies

Table 3.1 Selected research on code-switching in digital communication (adapted from Androutsopoulos, 2013b: 674)

Publication	*CMC Platforms/ Media*	*CS Languages*	*Participants/ Social Settings*	*Methods Quantitative (QN), Qualitative (QL), Mixed*	*Other Concepts and Themes*
McClure (2001)	Mailing lists	English, Assyrian	Ethnic minority	QL	Language maintenance
Georgakopoulou (1997, 2004, 2011)	Email	Greek, English	Friends	QL	Identities, footings, self-presentation
Lam (2004)	Chatroom	Romanized Cantonese, English	Immigrants, ESL students in the US	QL	Literacy
Hinrichs (2006)	Email	Jamaican Creole, English	University students, Jamaican diaspora	Mixed	Functions, identities
Siebenhaar (2006)	IRC, web chat	Swiss German, standard German	Youth	Mixed	Code choice, language variation
Androutsopoulos (2006, 2007)	Forums	German/Greek, Persian, Hindi, Arabic	Ethnic minority	Mixed	Identities
Chen (2007)	Bulletin board system	Mandarin Chinese (L1), English, Taiwanese	Adolescents	QL	Discourse functions
Lee (2007b)	Email, ICQ	Chinese (Cantonese), English	University students	Mixed	Effect of synchronicity
Su (2007)	BBS	Chinese (Mandarin), Taiwanese, English	College students	QL	Indexicality, playfulness
Deumert and Masinyana (2008)	SMS	isiXhosa, English	Young adults	Mixed	Homogenization
Warschauer et al. (2007)	Email	English, Egyptian Arabic	Young professionals	Mixed	Language choice
Huang (2009)	BBS, email	Chinese (Mandarin), English, Taiwanese	Chinese-English bilingual university students	QN	Script-switching

Leppänen et al. (2009, 2011)	Blogs, forums, fan fiction	Finnish, English	Young people	QL	Translocality, identities
Spilioti (2009)	SMS	Standard Greek, Greeklish, English	Youth	QL	CS vs. borrowing
Vandekerckhove and Nobels (2010)	MSN	West Flemish, Dutch	Teenagers	Mixed	Code eclecticism
Lexander (2012)	SMS	Wolof/Pulaar, French	Students in Senegal	QL	Literacy practices
Seargeant et al. (2012)	Facebook	Thai, English	Young adults, Thai native speakers	QL	Addressivity/Audience design
Androutsopoulos (2013a)	Facebook	Greek, German, English	German-Greek secondary school students	QL (online ethnography)	Networked multilingualism
Kytölä (2013)	Forum	Finnish, English	Football fans	QL	Metapragmatics
Bali et al. (2014)	Facebook (pages)	English, Hindi	English-Hindi bilinguals	QN	Natural language processing
Halim and Maros (2014)	Facebook (status updates)	Malay, English	Bilingual adults, English teachers	QN	Discourse functions
Jaworska (2014)	Forums	German, English	English-speaking German expatriates in Britain	QL	Language play, networked multilingualism, translanguaging, poly-/metro-lingualism
Themistocleous (2015)	IRC	Cypriot, standard Greek	Greek-Cypriots IRC users	QL	Identities
Thorne and Ivković (2015)	YouTube	Multiple	YouTube commenters	QL	Linguistic landscape, pluralingualism
Zhang (2012, 2015)	Douban, Weibo, Youku	Chinese (Mandarin), English	Commenters on government microblog	QL	Identities, language play

vary in terms of CMC platforms, languages on which the participants draw, participants' social backgrounds, as well as research methods.

In terms of platforms, researchers tend to study the popular ones at the time of writing so as to make timely contributions to the field. The studies in Androutspoulos's list also fall neatly into distinctive categories of *synchronous* or real time CMCs, such as IRC (Siebenhaar, 2006) and IM (Lee, 2007c), and *asynchronous* CMCs, such as email (Georgakopoulou, 1997) and discussion forums (Androutsopoulos, 2006). With the advent of social media and Web 2.0 technologies, dichotomies such as asynchronous versus synchronous and even public versus private have broken down. This is especially true of social media, which offer an assemblage of communication modes and spaces on the same website. For example, Facebook is a "collocation of online spaces" (Lankshear and Knobel, 2008) including mail and chat; it may be public or private depending on the privacy settings made by individual users. Since 2010, social media, especially Facebook and Twitter, have become popular research sites for CS in CMC studies (e.g. Seargeant et al., 2012; Androutsopoulos, 2013a). Studies on CS in social media and multimedia texts are still limited, as also noted in Androutsopoulos (2013b). Increasingly, however, there have been studies of CS in Facebook wall posts (Bali et al., 2014) and YouTube comments (Thorne and Ivković, 2015).

Numerous linguistic codes have been covered in previous studies. These include switching between Assyrian and English in a public mailing list (McClure, 2001), between Greek and English in emails (Georgakopoulou, 1997), between Finnish and English in blogs (Leppänen, 2007), between Wolof/Pulaar and French in SMS (Lexander, 2012), and between Swiss German dialects and standard German in IRC (Siebenhaar, 2006). While most of these examine CS on one CMC platform, there also exist comparative studies. For example, I have carried out a comparative analysis of Cantonese-English code-mixing practice in email and ICQ (Lee, 2007b) and noted that the participants code-mixed significantly more in ICQ than they did in emails. I have explained this in terms of synchronicity: To the participants, the mode of ICQ real-time interaction more or less resembles face-to-face conversations, where participants are already used to the practice of switching between languages. This practice seems to carry over to the writing of Facebook status updates (Lee, 2011), where Cantonese–English bilingual users rarely use mixed code; in that context, status updates are asynchronous and thus the need to code-switch is lower.

As far as participants are concerned, as with many studies of CMC, the "typical" group – young people or students – still dominates this body of research. For example, the user category of university students is explored in numerous studies, such as Hinrichs (2006), Lee (2007a, b), and Su (2007), while the broader category of young people or young adults is pursued in Chen (2007), Deumert and Masinyana (2008), and Spilioti (2009), which may also include student participants. Other studies are less specific about the demographics of their research participants. The labels "migrants" and "ethnic minority" are used to refer to different user groups. The former simply refers to groups of people who have moved overseas and do not necessarily switch between a dominant and minority languages (e.g. Lam, 2004), while the latter, borrowing Androutsopoulos's (2013b: 675) interpretation, refers to "situations in which an immigrant or diasporic group uses a minority language and a majority language". McClure's (2001) study of switching between Assyrian and English is a case in point. Other authors are more specific about their participants' occupations and interests, as in Kytölä's (2013) research on CS between Finnish and English among football fans in Finland and the Egyptian "young professionals" surveyed in Warschauer et al. (2007).

Methodologically, depending on the research questions of individual studies, qualitative and quantitative approaches and a mixture of both (mixed methods) have been adopted. Qualitative methods adopted in online code-switching include discourse analysis, interviews, participant observation, and ethnographic approaches. For example, Themistocleous (2015) carried out online participant observation, while Kytölä (2013) adopted mostly discourse analysis with an ethnographic style, or what Androutsopoulos (2008) refers to as *discourse-centered online ethnography* (DCOE). This means that while paying close attention to the textual data, the researcher also elicits insider perspectives on their language use through interviews or other forms of direct contact with the participants (see Chapter 8). Quantitative studies, on the other hand, are concerned with frequency and distribution of individual languages and instances of CS in a large pool of data. These range from a basic frequency count of the functions of CS (e.g. Halim and Maros, 2014) to corpus analysis of a large textual database (Bali et al., 2014). Mixed-methods research includes studies that make use of both approaches. One of the most common types is studies that combine frequency count with interviews, such as Deumert and Masinyana (2008). The overall trend suggests a shift to largely qualitative or mixed-methods approaches, while pure quantitative research seems to be becoming less common, or at least it is often complemented by a wider range of data sources.

In addition to identifying instances of CS in their CMC data, researchers of CS in CMC have become interested in understanding the discourse functions and motivations behind CS in their research contexts. For example, a number of studies point to the significance of CS as a resource for identity marking and self-positioning for multilingual participants (e.g. Lam, 2004; Leppänen and Peuronen, 2012; Themistocleous, 2015; see also Chapter 4 for identity performance in multilingual CMC). The rest of the chapter examines CS in CMC from both structural and functional perspectives.

STRUCTURAL PATTERNS OF CODE-SWITCHING ONLINE

As with spoken code-switching, CS in computer-mediated writing may occur at inter- or intrasentential levels. A number of studies begin with identifying and describing structural patterns of their CS data according to existing frameworks of CS structures in spoken discourse. Auer's (1998, 1999) broad distinction between insertional and alternational switches, for instance, has been adopted by numerous CMC researchers to describe the patterns of their CS data (e.g. Androutsopoulos, 2006; Hinrichs, 2006; Leppänen, 2007).

Insertional code-switching is a form of CS where one language or code is the base language of the utterance, providing the grammatical frame into which another language, or *embedded language*, is inserted (Auer, 1998). In email messages written by Jamaican–English bilinguals, lexical items in Jamaican Creole (Patois) are often inserted into an English-based email, as shown in Example 3.3

Example 3.3 Insertional CS: Jamaican Creole (underlined) inserted into an English-based email

By the way, nutten neva gwaan wid Shayzan *[Translation: Nothing happened with Shayzan]*. W went to another party afterwards and didn't leave until after 4 or so. She decided to go home after that. There will be other times however. . . . .hahahaha

Henry

(Hinrichs, 2006: 51)

Alternational code-switching is a process of CS in which two languages alternate within a discourse. There is a "true switch from one language to the other, from grammar to lexicon" (Muysken, 1997:361). Auer (1999) notes that in alternational switching, there tend to be preferences for "other-languageness" or using one language at a time. Leppänen (2007) further builds on Auer's model in her study of Finnish-English fan fiction and notes that, in fan fiction, alternational switching involves a text being written consistently in two languages to the extent that it is difficult to identify which the dominant language is, and the participants are expected to be proficient in both languages (Leppänen, 2007). The fan fiction excerpt in Example 3.4 is based on the American TV series *The Invisible Man.* In this segment, the author systematically switches and alternates between Finnish (translation in brackets) and English. The writer also uses one language at a time and there is no clear sign of one language standing out as the matrix.

Example 3.4 Alternational CS in fan fiction

"How are you?" Claire kysyy hymyillen. [Claire asks me, smiling.] . . .

Hobbes mulkoilee mua seisoen vielä ovella. [Still standing by the door, Hobbes eyes me crossly.]

"What you're smirking at over there, missy?"

"Nothing."

Mä katson muihin huoneessa olijoihin epävarmasti ja kerään rohkeutta ennen kuin avaan jälleen suuni. [I'm looking at the other people in the room, feeling unsecure, brace myself before opening my mouth again.]

"Uh, now, there's something you should know about me. . . . This",

Mä viittaan yleisesti ympärilleni, [I point to my surroundings,]

"is not really my reality. You don't exist in my reality. You're just a TV show in my reality." Hobbes hymähtä. [Hobbes gives a snort of contempt.]

"A TV show, eh? Where are the cameras?"

(Leppänen, 2007: 161)

Structurally, there is clear evidence of conventional alternational code-switching in that no matrix language can be identified. From a functional perspective, however, the Finnish and the English texts appear separated. That is, English is reserved for the dialogue only, while Finnish is for the narrative. In the context of fan fiction writing, this can be interpreted as a "narrative strategy" (Leppänen and Peuronen, 2012) – English is the language that represents the fictional character in the TV series, while Finnish is the language of the narrator's "authentic" Finnish identity (see Chapter 4 for a discussion of multilingual identity online). In so doing, the fan fiction writer creates a world in which the narrator enters the fictional world of the American TV series, where all characters speak English only. There are similar examples of alternational CS in Leppänen's study of fan fiction by Finnish young people (Leppänen, 2007; Leppänen and Peuronen, 2012).

Intraword switching occurs when the writer combines elements of two or more languages into one lexical item. For example, the English progressive aspect suffix *-ing* is often inserted in Chinese-based online discourse. An example is the post "荷包縮水ing.", found on a Taiwan-based student forums (Chen, 2007), which literally means "a wallet is shrinking" and can be translated into "[someone] is becoming poorer". This example of switching within a word also involves the process of script-switching between alphabetic writing and character writing.

For research on multilingual CMC users who are used to code-switching as an unmarked choice in their offline lives, the patterns of code-switching in their online discourse may strongly or partially resemble patterns of unmarked CS in spoken contexts (e.g. Deumert and Masinyana, 2008). In particular, in casual CMC contexts, one of the key motivations is to build solidarity, and a way of achieving that is to use an existing linguistic style that is familiar to all participants. As CS is an unmarked spoken feature in many bilingual communities, bilingual CMC participants would inevitably use code-switching as a resource for informal communication. This is also consistent with what my Hong Kong bilingual participants claimed in various interviews. Many of my participants, when asked why they used both written Cantonese and English in their CMC, saw their CS as naturally occurring or said that they wanted their chat messages to sound friendlier because CS was exactly what they would do in everyday, especially informal, talk.

However, online participants do not simply transfer their CS practices in spoken interaction to their online discourse. This is evidenced by some of my participants' rejections to some of their own uses of CS online. When shown their own examples of CS in IM, some participants appeared surprised or even expressed resistance to their code-switching examples in the data. Comments such as "That [CS example] sounds so strange!" or "I would not say this in English!" were not uncommon in the interviews. What makes code-switching research in CMC more intriguing is when multilingual writers do *not* follow conventions and expectations in spoken CS, whether unconsciously or deliberately. New "styles" of switching have been reported on in various studies of CMC. For people from social contexts where CS is unmarked or expected, both "mainstream" and "nonmainstream" styles have been identified. According to Chen (2005), in spoken interaction, mainstream styles of CS are features that have been identified extensively in previous literature. For example, in Hong Kong, CS is mainly *insertional* at the intrasentential level, and Cantonese is expected to be the base language into which English words are inserted. Nonmainstream style, by contrast, deviates from the expected norms of CS patterns that have been identified in previous work. The email message in Example 3.5 exhibits a stylized, nonmainstream pattern of Hong Kong–based CS.

Example 3.5 Stylized nonmainstream code-switching in an email

Dear DD,
> Hee hee . . . dunno why I always like to send u mails ar! Part is
>becoz I wanna keep contact with u la! Another reason is I am having
>"sky and land" lessons today!

This is an email sent to a close friend by a Hong Kong Chinese university student. The base language is English, into which Romanized Cantonese particles (e.g. *ar, la*) are inserted. The transliteration of Cantonese expression (sky and land lessons 天地堂), as already explained in Chapter 2, demonstrates the writer's creativity and is deliberately playful. This pattern of CS would have been considered awkward should it appear in offline, spoken interaction. As mentioned, when Hong Kong Cantonese–English bilinguals code-switch in speech, Cantonese is the main language on which the grammar of the code-mixed sentence is based. Example 3.5, however, shows the opposite: The dominant language is English, while Cantonese discourse particles in Romanized form are inserted to the end of the English sentences. This practice of inserting Romanized particles is found to be extremely common in online communication among Hong Kong people (as discussed in James, 2001; Lee, 2007a).

Apart from using a different base language from the expected norm of CS in speech, words that are not normally inserted in spoken CS situations may be found in online CS, as illustrated in Example 3.6.

Example 3.6 An IM message illustrating nonmainstream code-switching

Billy: 喂, Today 無 節目 咩? *(Translation: Hey, don't you have any plans for today?)*

Although structurally this example qualifies as a case of insertional switch, according to the literature, as well as my knowledge of Cantonese-English CS in Hong Kong, *today* (or any adverbs in English, such as the case of *only* discussed in Barton and Lee, 2013: 50–51) is not a word that would be added to a spoken CS context. Under normal circumstances, the question in Example 3.6 would have been asked in Cantonese only. A similar phenomenon is also identified in Spanish–English email (Negrón Goldbarg, 2009), in that both English and Spanish may serve as the base language of an email, depending on various situational factors.

For speakers who are not used to CS in their spoken interaction, CS in online communication tends to be more stylized, playful, and creative (Zhang, 2012, 2015). In mainland China, Chinese–English CS is not uncommon. However, compared to the situation in Hong Kong, CS in everyday spoken conversation is a less frequently occurring and much more recent phenomenon. In the literature of CS in China, CS occurs mostly in specific genres, such as classroom interaction, news reporting, advertising, and more recently the internet (Zhang, 2012, 2015), such as in Example 3.2 at the beginning of the chapter, repeated here:

> 多云中午前后转阴有雨,局部大雨,26–32度: #早安上海# Everybody, it's holiday today. Are you ready? today 又是 Sunday, 我好想 day day 都非 hot day,Where good where cool 我们 where stay, NO 冷 NO热, Love 谁谁. It will be rainy! I'm sorry. 冷空气南下与暖湿气流交汇,雨水难逃好乐 day
>
> *(Translation: Cloudy to rain around noon, isolated heavy rain, 26–32 degrees: #Good morning Shanghai# Everybody, it's holiday today. Are you ready? Today is Sunday again, I do not want to have hot day everyday; We stay at wherever is good and cool, NO cold NO hot, Love who [who cares]. It will be rainy! I'm sorry. Cold air will land in the south and meet warm and moist airflow, hard to avoid rainfall on holiday.)*
>
> (Zhang, 2015: 243)

The beginning of the post mimics a conventional weather reporting style (Cloudy to rainy . . . 26–32 degrees), followed by a playful adaptation of the original lyrics of a popular Chinese song called 好乐day. For example, the *Sunday* in "Today 又是 Sunday" (Today is Sunday again) in the original lyrics now becomes *Saturday* in the post; "no work, no learn" in the lyrics becomes NO 冷 NO热 (no cold, no hot, meaning it's neither hot nor cold). It is almost impossible to tell what the base language is and whether the switches are insertions or alternations. As it turns out, the structure of the post makes intertextual references to the song lyrics. 好乐day also exemplifies the switching between Chinese characters and the Roman script within one word. It would be inappropriate to label the example as either Chinese or English. Playful multilingual writing such as in Example 3.2 has been taken up extensively by official Chinese microblogs. An earlier study by Zhang (2012) has reported similar forms of written CS on the official Weibo mircoblog site of a regional police department. Such playful style of written CS, according to Zhang (2015: 243), has been "warmly received by many netizens". However, when taken out of its CMC context, this message would be unexpected or even unacceptable. I conducted a short email interview with Xiao, a mainland Chinese postgraduate student, who is proficient in English and Mandarin Chinese, for her opinions about this specific example. Her immediate response was that this post looks "VERY weird" and she added that:

> [this kind of switching] should never appear in a natural context (either oral or written) when communicating with other people. If someone use[s] codemixing so much in a single sentence, I would think "doesn't he/she know how to speak normally?" To me, this kind of expression is only acceptable when people are intentionally codemixing, or talking about codemixing.
>
> (Xiao, 2015, personal communication)

Zhang (2015: 244), however, argues that this form of "multilingual meme-ing", as she puts it, illustrates the creative function rather than just communicative function of language. It also gives rise to a hybridized discourse style in which "bits and pieces of English(es) have been manipulated as cultural resources in building the image of the Shanghai Municipal Government as a modern, open, and vigorous government that identifies with urban youth and 'advances with times'". It seems reasonable to further interpret instances of code-switching in CMC in terms of their discourse functions.

DISCOURSE FUNCTIONS AND SOCIAL MOTIVATIONS OF ONLINE CODE-SWITCHING

In many cases of CS in online discourses, CS does serve to mark various immediate discourse functions as it does in speech. Summarizing existing research on CS in CMC, Androutsopoulos (2013b: 681) identifies the following functions in previous literature:

- switching for formulaic discourse purposes, including greetings, farewells, and good wishes;
- switching in order to perform culturally specific genres, such as poetry or joke telling;
- switching to convey reported speech (as opposed to the writer's own speech);
- switching with repetition of an utterance for emphatic purposes;
- switching to index one particular addressee, to respond to language choices by preceding contributions, or to challenge other participants' language choices;
- switching to contextualize a shift of topic or perspective, to distinguish between facts and opinion, information and affect, and so on;
- switching to mark what is being said as jocular or serious and to mitigate potential face-threatening acts, for example through humorous CS in a dispreferred response or a request;
- switching to or from the interlocutor's code to index consent or dissent, agreement and conflict, alignment and distancing, and so on.

An instance of CS in CMC may serve more than one of these listed purposes. For example, the insertional switch to English "is there no gravity in Indian brains?" at the end of Example 3.1 serves to emphasize the writer's emotional concluding remark as well as to distinguish itself from the previous argument in the post. Of course, not all instances of CS in CMC would fall neatly into these categories.

While these functions have been derived from studies of CS before social media, recent research has become interested in the functions of CS in convergent social network sites. For example, Halim and Maros (2014) identify a list of 11 discourse functions of CS on Facebook. Many of the functions are indeed in line with Androutsopoulos's categories, such as switching for reiteration, addressee specification, emphasis, and so on. Additional functions have been identified in the Malay–English CS posts on Facebook collected by Halim and Maros:

- *Switching for checking* is used to seek consensus, approval, and confirmation from the addressee. An example would be switching to a Malay tag question at the end of an English post, as in: "It's funny when I know that I know what you know . . . Kan?" (Right?)
- *Switching for availability* occurs when a word or expression is only available in a particular language. This often happens when English is the base language, to which a Malay expression is inserted when the entity described is not available in English. For example, in the post "celebrating my personal hijrah", *hijrah*, which literally means "to leave a bad way of life for a good way", does not have any equivalent in English. Inserting cultural-specific expressions is comparable to what some may consider "cultural borrowing" (see Myers-Scotton, 1997).
- *Switching for principle of economy* matches with the basic principle of human communication, in that language users tend to choose the shortest and easiest words to communicate. This is found to be a key motivation for spoken CS not only in Malay–English CS but also in other contexts such as Hong Kong Cantonese–English CS (Li, 2000).
- *Free-switching* is an additional category introduced by Halim and Maros to refer to CS that does not serve any specific pragmatic or discourse-related function. This is similar to the form of stylistically motivated switch in the hybrid Chinese–English weather message in Example 3.2.

In fact, many instances of CS in CMC are socially motivated instead of serving any of the mentioned discourse functions. For example, Georgakopoulou (1997) and Hinrichs (2006) have noted that code-switching in email messages is often motivated by the level of formality of the communication context as well as the relationships between participants. In a similar fashion, in Taiwan-based CMC, script-switching between standard Chinese characters and Zhuyin (Mandarin phonetic writing system) reveals the user's desire to adopt an informal style and a jocular tone. (Su, 2007; Huang, 2009).

In the age of social media, CS becomes a more complex topic when different forms of sharing and textual formats are available on one single social network site. On Twitter, for example, multilingual users have been found to write a post in one language, which is then followed by hashtags in another language, as in Examples 3.7 and 3.8.

Example 3.7

Eating an apple for lunch while everyone around me eats cheeseburgers and fries. #yoquiero

Example 3.8

Jetzt gibt's was vernunftiges zum essen! #salad #turkey #lunch #healthy #healthylifestyle #loveit

(Jurgens et al., 2014: 51)

The tweet in Example 3.7 starts with an English-only post, followed by a hashtag in Spanish which literally means "I want" or "I like". In Example 3.8, the writer tweets in German ("Now there is something sensible to eat") with English-only hashtags that describe the meal and the writer's feelings about it. Although the motivation behind mixed-language tweets still remains uncertain, CS involving hashtags presents an interesting avenue for further research into multilingualism in social media. First, it may be debatable as to whether these examples count as switching or whether they simply exemplify the coexistence of two languages in the same post. This depends largely on whether the researcher considers the first part of the post and the hashtags that follow as one coherent text or as separate texts. Second, researchers of Twitter and English hashtags tend to understand hashtags as serving specific communicative functions in their own right (Small, 2011; Page, 2012). For one thing, that hashtags are hyperlinked gives rise to "ambient affiliation" and facilitates "searchable talk" in social media (Zappavigna, 2015).

A creative form of code-switching that is often found on Instagram is switching that occurs within a hashtag. For example, *#守MK* (protect MK, abbreviation of Mongkok, one of the occupied sites), *#fuck藍絲帶* (fuck blue-ribboners, supporters of the government and police) were amongst the hashtags on Instagram during the Hong Kong umbrella movement. A possible explanation for these instances of intrahashtag CS is that these bilingual hashtags allow the poster to visually display their affiliation to Hong Kong's bilingual culture, thus expressing their affection and support for their home city. These code-mixed hashtags may also serve as an important resource for expressing resistance to the political hegemony of mainland China, where Chinese-only discourses prevail, thus asserting the poster's political stance during the movement (Lee and Chau, 2015a). Further, because Instagram is essentially an image-based network, the meaning of hashtags needs to be interpreted in light of the content of the image posted (see Barton, 2015a on tagging on Flickr). It becomes clear that structural or functional analyses of CS ought to be complemented by qualitative data to reveal the social meanings of CS from the language user's perspective. More recent approaches to code alternations, as discussed briefly in Chapter 1, have moved beyond structural analyses of CS (or even beyond the term CS altogether) to exploring CS as a form of everyday practice for bilinguals or multilingual speakers (see Chapter 1).

RETHINKING CODE CHOICE AND CODE-SWITCHING IN THE DIGITAL AGE

Bringing together the discussion of language choice in the previous chapter and code-switching in this chapter, the traditional notions of language choice and CS are still useful in that they offer a starting point for understanding what happens to forms and functions of multilingualism when they are mediated by digital technology. Digital communication clearly offers a new and major domain for multilingual writing in the 21st century. As shown in Chapter 2 and in this chapter, while some patterns and functions of linguistic resources and code alternations resemble those in spoken contexts, new resources, patterns, and processes of multilingualism have been identified on the internet.

However, existing frameworks and classification systems are insufficient for understanding online multilingualism. One possible issue of relying solely on pre-existing categories is researchers *decontextualizing* multilingual writing from its actual situation of use (Androutsopoulos, 2013b). This problem can be minimized by complementing language or feature identification with ethnographic methods including interviews and participant observation (See Page et al., 2014). Sebba (2012) emphasizes the importance of moving beyond code-switching to understanding mixed-language texts from a literacy practice perspective. This is because instances of multilingual writing are situated within the broader linguistic and social context of the participants (Lee, 2007a). Assigning "neat" categories of multilingual resources or code-switching patterns to CMC data may also "prevent us from dealing with language production which cannot be ascribed to any individual 'language'" and "will inevitably simplify the range of resources employed by speakers" (Jørgensen et al., 2011: 6).

We need new ways of thinking about multilingualism in the digital age. One emerging trend in recent years is to take up the concepts of *superdiversity* and *mobilities* in research on digital media (e.g. Androutsopoulos and Juffermans, 2014). Under the superdiversity framing, linguistic phenomena that used to be understood in terms of strict categories or patterns of language have now become fluid *digital practices* which are constantly changing and flowing. The analytical focus has also shifted from structural patterns to people, their contexts, and associated social meanings. In this emerging body of work, the close relationship between digital language practices and identity construction is often foregrounded. For example, Juffermans et al. (2014) examine how Dutch-Chinese young forum participants associate with and distance themselves from "Chineseness" through participating in the forum discussions mostly in Dutch. This is interesting in that online participants may engage in multilingualism without actually using multiple languages. Terms such as *languaging*, *translingual practices*, and *polylanguaging* (as discussed in Chapter 1) have been adopted to describe the blending of linguistic codes that serves specific social functions. The case of Chinese–English weather messages (Zhang, 2015) is not simply about being playful. What's behind playfulness is the writer's intention to speak to the younger web generation as well as the reader's acceptance or nonacceptance of such playful representations of a formal genre (i.e. weather reports). These issues are all closely related to identity construction and self-presentation online. The next chapter is devoted to the relation between multilingualism and identity in the online world.

4

MULTILINGUAL PRACTICES AND IDENTITIES ONLINE

Overview

- Multilingualism and identities online
- Performing glocal identities in translocal online spaces
- Hashtag politics: Asserting local voices through Instagram hashtags
- Language play and identities in social media
- Negotiating identities in online teaching and learning spaces
- Doing identity work with multilingual resources online

Everything we write online reveals something about ourselves. Even in the most "anonymous" web environment, we easily leave traces of who we are by writing about certain types of content, choosing a particular screenname, or using certain linguistic codes to write a message. Returning to my technobiography discussed at the beginning of Chapter 1, it is clear that I do not draw on the same set of linguistic resources across different online platforms. Even on the same site, I deploy my resources according to my audiences, the content of the post, and other contextual factors. On Facebook, for example, sometimes I post a status update in English only, especially when talking about some serious or work-related topics. However, when I post about something more personal, such as sharing photos of a family gathering, I may write in Cantonese or I may mix Cantonese and English.

At other times, I may comment on my close friends' wall posts in Romanized Cantonese to ensure that few people can "decode" our messages. My language choice on Facebook not only reflects the ways in which I design my digital writing according to my audience groups, but it is also shaped by *who I want to be* when I post my messages, such as a friend, a colleague, a family member, or a playful or serious person. In other words, as with face-to-face communication, language choice and code-switching online have always been tied closely to the topic of *identity*. As will become clear in the discussions in this chapter, languages serve as powerful resources for web participants to assert and position themselves as different kinds of people and perform different aspects of identity. This chapter is concerned with the relationship between multilingual resources and identity performances in online spaces.

MULTILINGUALISM AND IDENTITIES ONLINE

Though not always discussed explicitly, much existing research on code choice and code-switching in online communication leads to a common theme: that the negotiation of language choice and alternation between linguistic codes serve as important resources for self-presentation and identity construction. Identity is a multifaceted concept, and the conceptualization of it often goes beyond categorizing human beings into fixed properties. Although certain aspects of identity are considered more static than others, such as age, gender, and nationality, other aspects such as hobbies, interests, and social networks are more subject to change with time. There are also aspects of identities that are shaped by social domains (e.g. work, family, and education) and relationships (e.g. friends, colleagues, and family). In view of the multiplicity of identity, this book adopts the plural form *identities*. This conceptualization highlights the fact that identities are always open to reappropriation, recontextualization, and transformation.

The term *identities* needs to be defined before examining its role in online multilingualism. Identities are socially constructed and may change as a result of various factors, such as the context of interaction or the ways interlocutors interpret the identities being projected. In social interaction, people are constantly making decisions about whether they wish to express or reveal certain aspects of their identities. This is also in line with Erving Goffman's (1959) analogy of identity as a theatrical performance. According to Goffman, identities are like masks that can be worn and taken off by social actors in different communicative situations. The performance and reappropriation of identities are what Goffman refers to as *impression management*. In impression management, there is often tension between the different kinds of self that people project (Higgins, 1987; DiMicco and Millen, 2007). Sometimes the actual self (aspects of identity that one possesses) is revealed; at other times, people may want to project their ideal self (what one wants to possess) and their ought self (what one should possess) (Higgins, 1987). Likewise, identities in digital media are not just about who we are, but also who we want to be to others, and how others see us or expect us to be. In online participation, people do not behave as single, self-contained beings, but as networked individuals (Wellman, 2001). In other words, identity work is carried out collaboratively among web participants.

Conceptualizing identities as a dynamic concept is useful in understanding language and identities on the internet. Identity work done through language is evident in all kinds of computer-mediated discourse, from a screenname in Internet Relay Chat (IRC) to commenting in social media. For example, in IRC, playful and carefully designed nicknames are an important means to draw attention so as to initiate a new conversation (Bechar-Israeli, 1995; Lindholm, 2013). This is because associative meaning is usually embedded in a nickname to signal aspects of one's identity (e.g. "Blondie" is suggestive of the participant's hair color).

Of relevance to this book is how multiple languages are deployed to perform identities. Much identity work in multilingual CMC aims to establish relationships with others. In Egypt, the choice between English and colloquial Arabic writing is a signal of formality. Warschauer et al. (2007) noted that English was reserved for formal email while Romanized Arabic was more prominent in informal email and chat. In a similar way, switching between English and Greek in email served to enhance solidarity as well as indicate professional in-group membership (Georgakopoulou, 1997). In Senegal, French, rather than African languages, was found to be the dominant language of romantic text messages among lovers (Lexander, 2012). These studies suggest that traditionally, formal, and professional online communication, such as work-related email, is associated more with a dominant language such as English, while a local language is reserved for informal and interpersonal interaction online, such as chatting.

Turning to the age of Web 2.0 social media, the technological affordances of digital media provide opportunities for users to draw upon a much wider range of multilingual and multimodal resources for self-presentation. A characteristic of social media platforms such as Facebook and Twitter is that they provide users with multimodal self-presentation spaces that combine images, videos, written and spoken language, and more. Such spaces can be referred to as *spectacles*, that is, "multimedia content that is produced outside media institutions and uploaded, displayed, and discussed on media-sharing websites" (Androutsopoulos, 2010: 203). Video blogs (vlogs) or monologues on YouTube, for example, are often accompanied by written information, including titles and tags. These resources combine to form a lens or a curation platform (Snyder, 2015) through which the audience can extract information about the video poster's identities. One interesting dimension of identities in social media is how uploaded contents often draw attention to the self and are then shared *publicly*. Examples of such public presentations of oneself include the countless first-person vlogs found on YouTube, as well as selfies on Tumblr, Instagram, and other photo-sharing platforms (Tiidenberg and Cruz, 2015). In fact, some of the most popular hashtags on Instagram draw attention to oneself, such as #me (Webstagram, 2014). These phenomena clearly illustrate what Crandall (2007) refers to as *presentational culture*, in which people constantly pay attention to their own image or aspects of identity to be projected to a wide range of audiences. The sections that follow focus specifically on cases of identity performance through multilingual resources.

PERFORMING GLOCAL IDENTITIES IN TRANSLOCAL ONLINE SPACES

In Chapter 1, I argued that the internet has become an important element in the age of supermobility and diversity. Digital media have afforded important *translocal* and *transnational* spaces for people to engage in new kinds of communicative practices and linguistic resources. These new practices and resources also enable online participants to assert new kinds of local and global identities. Before the age of social media, script-switching was already noted among Greek web users who employed "Greeklish", or Romanized Greek, in discussion lists and SMS (e.g. Koutsogiannis and Mitsikopoulou, 2003). This practice is considered a way for web participants to strike a balance between their linguistic practices in the increasingly global world and their attachments to their local linguistic identities. This process is often referred to as *glocalization*, which is people's "dynamic negotiation between the global and the local, with the local appropriating elements of the global that it finds useful, at the same time employing strategies to retain its identity" (Koutsogiannis and Mitsikopoulou, 2007: 143).

Web users' concurrent use of English and other languages in order to project a cosmopolitan identity is especially evident in fan fiction produced by young people around the globe. Fanfiction.net is a website where fans of creative works (e.g. novels, mangas, TV series, movies) rewrite and publish their versions of the original works. The majority of the works posted on the site are written in English; the site, however, attracts writers who claim to be second language users of English. Fan fiction is also a form of collaborative writing that serves as "a performative means for establishing and negotiating identity and shared culture, and an index of a particular lifestyle and values" (Leppänen et al., 2009: 1090). In Finland, young fan fiction writers were found to use English alongside other languages to index translingual and translocal experiences of the writers. Example 4.1 is taken from a piece written by a young Finnish fan fiction writer, Afeni. The story is based on the Japanese manga *Naruto*.

Example 4.1 Extract of code-mixed fan fiction by a Finnish writer

"Aren't you envious, are you, ototo-san?" Itachi said a playfully tone in his voice.
He stood there laughing
I felt the kunai in my hand and he laughed no more

Since I was still wearing my day clothes I reached to my kunai bag and grabbed the kunai. . . .

(Leppänen et al., 2009: 1091)

While the post is written largely in English, there are occasional insertions of Japanese proper nouns and lexical items adapted from the original story. *Ototo-san* and *Itachi* are names of the characters in the story, while *kunai* means "knife" in Japanese. According to Afeni, writing in English was sometimes a natural choice, and she was aware that English allowed her to reach a wider readership. This in a way reflects Afeni's perceived status of English as a global language, as discussed in Chapter 2. The addition of Japanese words and cultural elements enhanced the degree of authenticity of the conversations between the characters, whose first language is Japanese in the original manga or anime. In addition, the Japanese words in the fan fiction indexed Afeni's insider knowledge of Asian culture and languages, which is highly valued in the communities of fan fiction writing. The symbolic uses of an Asian language also granted her considerable authority in a highly multicultural fan fiction world.

A similar pattern was also observed in Black's (2009) study of three Asian fan fiction writers. As with the Finnish fan fiction writers, the three young participants in Black's study drew on knowledge of their multiple linguistic and cultural resources in their English-based fan fiction texts. In the fan fiction texts in Black (2009), the three focal participants all started their texts by positioning themselves as rather incompetent writers of English. For example, one of the writers, Grace, wrote: "English is not my FIRST LANGUAGE" and admitted to her "poor English" in the header information of her fan fiction page. These seemingly self-deprecating comments have been found to be a salient practice of gaining support from other fan fiction writers and other social media participants (Lee, 2012; see also discussion in Chapter 5). In addition to English, the writers all drew on other languages in their texts. For example, like Afeni, they inserted Romanized Japanese phrases and lyrics in the header information as well as within the main body of their fan fiction. One of the focal participants, Nanako, drew on both English and Mandarin Chinese, her first language, in her fan fiction writing. Example 4.2 is part of her text.

Example 4.2 **Nanako's *Crazy Love Letters***

Meiling turned to Syaoran and grumbled in Chinese. "Dan shi, Xiolang, wo xiang he ta shuo ji ju hua" (*Translation: But, Xiaolang, I just wanna talk to her.*)

(Black, 2009)

In this short example, the main characters' names, as well as the quotation, are written in Romanized Chinese, or *Pinyin*. As Black argues, this linguistic practice, on the one hand, immediately indexes Nanako's Chinese identity; on the other hand, the switching between Chinese and English demonstrates her linguistic abilities as a multilingual young person in the fan fiction community. These examples of fan fiction writing practices illustrate the ways in which the fan fiction site has been transformed into a translocal affinity space (Leppänen and Peuronen, 2012), that is, a transborder contact zone in which fan fiction writers from around the world can project their multicultural and cosmopolitan identities through reading and writing mixed-language texts.

Web 2.0 social media have brought translocality to the forefront. As with fan fiction, the photo site Flickr is a globalized site where both verbal and visual contents are posted by users and are then shared globally to readers from all parts of the world. As mentioned in Chapter 2, a considerable proportion of the Chinese and Spanish participants in Barton and Lee's (2012, 2013) research considered English to be the universal language on Flickr, and its use helped them reach a wider audience. For example, a Spanish user, "Carolink", said she had never participated in a global online network before, and yet she gradually identified herself as a bilingual global citizen. She explained in an interview that Spanish was "toọ limited for these Internet times. I do not leave Spanish, but I try to use English when I can". However, not every Flickr user is willing and ready to participate in English. In an online interview, a mainland Chinese participant, "sating", challenged the notion of globalization as well as the globally recognized status of English:

> If Flickr is a global website, as a Chinese, why must I use English, a language that I am not good at? Besides, most of my photos reflect the reality of China. So Chinese has to be the most suitable tool of communication.
>
> (sating, personal communication [original interview in Chinese])

This comment indicates both sating's rejection of English as a global language and her strong desire to assert her Chineseness on Flickr. To sating, Flickr should be a space for celebrating intercultural and cross-linguistic exchanges rather than one that limits users to one language. Chinese serves as a marker of her local identity because it can attract Chinese-speaking Flickr users within China and beyond. As with using Chinese in English fan fiction, Chinese on Flickr helps promote sating's cultural and linguistic background to the wider Flickr community.

The complex relation between the local and the global is also evident in the creation of Flickr screennames. For example, a Hong Kong–based Singaporean Chinese Flickr user, "Kristie遊牧民阿靜", explained her mixed language screenname as follows:

> I want a wider group of people to know me. Not that the Chinese won't know me if I call myself just Kristie but if I attached a more "graphic" Chinese word (that's how I always see the language), we can connect quicker and better. The name also says a lot about who I am in my whole darn life.
>
> (Kristie遊牧民阿靜, personal communication)

As mentioned in previous chapters, multilingual resources online often serve more symbolic than pragmatic functions. To Kristie遊牧民阿靜, the visual aspects of Chinese and Roman scripts in her screenname act as an important emblem of her cosmopolitan identities. To summarize, from Carolink's remark "I do not leave Spanish" to the Chinese users' maintenance of Chinese characters, multilingual practices on Flickr encourage glocal and heteroglossic literacy practices. These users wish to participate in the global online networks through English (even with limited knowledge of it) without giving up their existing local identities. This also demonstrates that the concurrent use of multiple scripts and languages does not necessarily reflect users' competence in the languages, nor does it automatically reveal one's ethnicity. In the digital world, traditional boundaries of localities and nationalities become less significant sociolinguistic variables. Rather, the choice of language is closely related to the extent to which participants intend to project themselves as *global* or *local* members.

HASHTAG POLITICS: ASSERTING LOCAL VOICES THROUGH INSTAGRAM HASHTAGS

As discussed in Chapter 2, English still dominates various globalized social media, at least from the users' perspective. On Flickr, many users who do not speak or write English in their everyday lives seem to be keen to use some English on the site. However, a different data collection method may reveal a rather different set of language ideologies in relation to identity performances. For the Flickr project, David Barton and I sampled the first 100 photo sites we came across regardless of content. If, however, we zoom in to posts about a particular social event, such as the umbrella movement mentioned in Chapter 2, new affordances of linguistic resources may be discovered. When using social media to express feelings and opinions about a social event, participants also select their language resources in ways that are different from writing posts for other purposes. In the following, I present a case study of multilingual hashtags posted on Instagram about the umbrella movement and explore the interplay between language choice for hashtags and the projection of Hongkongers' identities (Lee and Chau, 2015).

The umbrella movement, or umbrella revolution, in 2014 in Hong Kong was driven by a series of pro-democracy campaigns and protests where supporters demanded "real" universal suffrage in Hong Kong. Streets in different locations of Hong Kong were occupied by protesters as a result of ongoing conflicts between the protesters, the police, the Hong Kong SAR government, and the central Beijing government. As with many social movements around the world, the umbrella movement had a strong digital and social media presence. Among these sites is the relatively underresearched platform Instagram. Similar to Flickr, Instagram is essentially an image-based social network site. In addition to posting images, the poster may add optional *hashtags* – user-defined searchable keywords prefixed with the hash symbol (#). A hashtag may consist of just one word or a string of words written without spaces. The practice of adding hashtags is also common on Twitter, Tumblr, and Facebook.

In the study, 700 Instagram posts with the Chinese hashtag #雨傘運動 (umbrella movement) were collected and analyzed by their language choice. It was found that the majority of these posts were tagged bilingually in Chinese (including Cantonese) and English, such as the following set of hashtags. The following hashtags describe one image that portrays a crowd in action at the protest site in Central, Hong Kong. Of the 23 hashtags, only three are in English. The rest are either written in Cantonese (e.g. #全城撐學生, "the whole city supports the students") or standard written Chinese (e.g. #人在中環, "man in Central").

> #罷課不罷學 #罷課 #689 #佔中 #全城撐學生 #路姆西 #真普選 #佔領中環 #金鐘 #中環 #銅鑼灣 #旺角 #和平佔中 #雨傘運動 #加油香港人 #自己地方自己救 #自己香港自己救 #和平抗爭 #人在中環 #我係香港人 #scholarism #HongKongers #umbrellarevolution

The significant presence of Cantonese tags in the Instagram data contrasts with earlier findings about Flickr, on which about half of the Chinese users still preferred to describe and tag their photos in English only (Lee and Barton, 2011; Barton, 2015a). On Flickr, although Chinese tags were used alongside

English ones, Cantonese tags were almost absent in our research. The range of purposes of bilingual writing on Instagram is also quite different from what is observed on Flickr. Chinese is often a language "on display" on Flickr, meaning that it is for the global non-Chinese speaking community to see, so as to assert a cosmopolitan identity on a globalized social network. Similar functions are still observed on Instagram where hashtags create "ambient affiliation" (Zappavigna, 2015), that is an online community built around user-defined topics of interest among Instagrammers at both the local and global levels. However, during the movement, the use of Cantonese hashtags and traditional Chinese characters became a powerful linguistic practice for asserting a local voice and self-positioning. As one Hong Kong Chinese Instagram user, Vincent, said in an online interview:

> Chinese hashtags represent the hong kong spirit and enhance the identity recognition of hk ppl . . . you know we use traditional chinese not simplified language in China. Traditional chinese is the language of all hk ppl . . . in hong kong, we can still use Cantonese. This language represent hk ppl and hk culture.
>
> (Interview, unedited)

Vincent's repeated uses of "hk ppl" (Hong Kong people) and distancing himself from "simplified language in China" clearly reveal the ongoing conflicts between Hong Kong and Mainland China. This tension between Chinese languages and scripts is not new and arose from outside the internet. On the protest sites, the strong presence of Cantonese writing is largely attributed to Hong Kong people's desire to defend their identity as Hongkongers, as opposed to mainland Chinese (Guilford, 2014). This is also a result of the growing conflict between the Mandarin-speaking central government in Beijing and Cantonese–English bilingualism in Hong Kong. The de facto status of Cantonese and the traditional character writing system in Hong Kong have been upheld since the colonial era.

The difference between the two varieties of Chinese is reflected in the use of the name for the movement – the word for umbrella is 遮 *ze1* in Cantonese, but the movement is often referred to in standard Chinese as *jyu5 saan3* 雨傘 (umbrella) *wan6 dung6* 運動 (movement) simply because Hongkongers grew up learning standard Chinese in school. In view of this, the term *ze1daa2 wan6 dung6* 遮打運動 was invented as a local reference to the movement. 遮打, to the average Chinese speaker, would literally mean the name of a street in Central, Chater Road. To Hong Kong Cantonese speakers, however, 遮 and 打 have an additional layer of symbolic meaning, in which 遮 is *umbrella* in Cantonese and also means "to block", and 打 means "to attack" or "to hit", both vividly reminding local Hong Kong people of the happenings during the movement. Figure 4.1 is a graphic illustration of the two possible characters for *umbrella* by a group of Cantonese scholars at the University of Hong Kong. The image was produced for and shared on the HKU Canto Facebook page.

Another important finding is that, among the 100 distinct Cantonese hashtags collected from the 700 posts, some deliberately express insider knowledge which may be unfamiliar to speakers of other Chinese languages or to people from outside Hong Kong. For example, #乜乜乜 (literally "etc. etc. etc.")

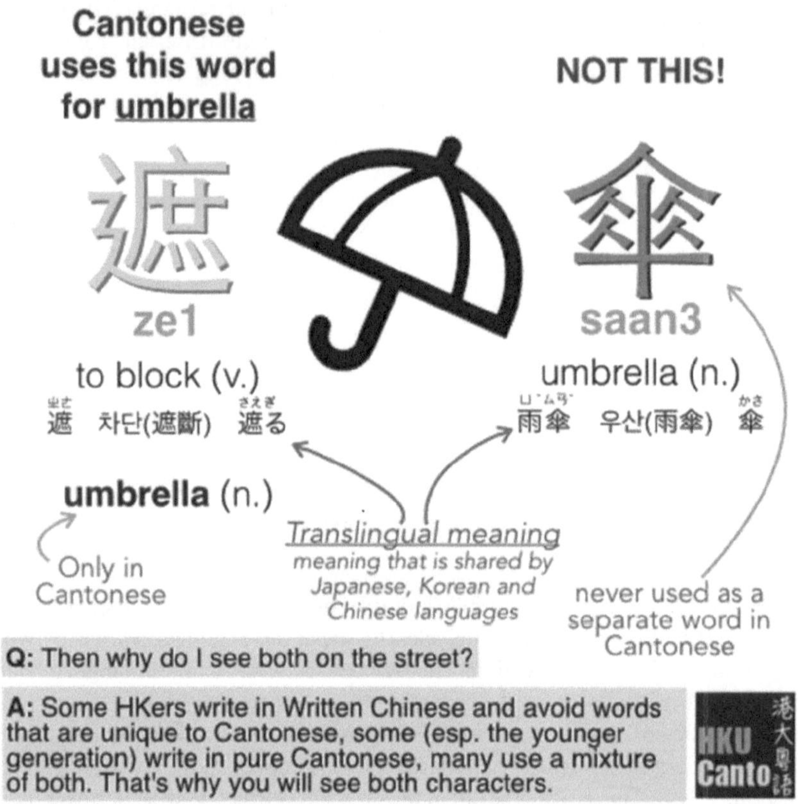

Figure 4.1 Cantonese and standard written Chinese of the word for *umbrella*

is a Cantonese expression that mocks those who are against the movement; it is an intertextual reference to a vague comment made by an opponent of the movement in a TV interview. Other Cantonese hashtags in the database include #撑學生 (support students) and #撑香港 (support Hong Kong), in which the verb 撑 means "to support" in Cantonese. Chinese–English code-mixed hashtags are also evident, as in #alexter萬歲 (long live Alexter, Alexter being one of the student leaders) and #守MK (defend MK, MK being the abbreviation for Mongkok, one of the protest sites). Again, these hashtags may be incomprehensible to non-Cantonese users, as Cantonese is deliberately taken up as a we-code by the Cantonese protesters.

In addition to Cantonese and mixed language writing, some other hashtagging practices have been adopted by supporters of the protest to perform their Hongkonger identities:

- *Claiming ownership*: A significant proportion of the hashtags collected express people's sense of belonging to Hong Kong and the movement. Chinese examples include #我係香港人 (I am Hongkonger) and #香港是我家 (Hong Kong is my home). In English, the first person possessive pronouns *my* and *our* allow participants to claim ownership of Hong Kong, such as #thisisourmoment, #fightforourfuture, #myhome.

- *Multilingual references to Hong Kong*: Of the 1,289 distinct hashtags identified, 120 (9.3%) made explicit reference to the main site of the umbrella movement 香港, 96 (7.4%) hashtags contained "hongkong", and 193 (15%) had the abbreviation "hk". Many hashtags denote places in Hong Kong, especially those associated with the protest sites, such as #MK and #金鐘. An interesting reference to Hong Kong is the number hashtag #852, which is the dialing code for Hong Kong. Because hashtags are hyperlinked and searchable, these location hashtags can be seen a sign of Hong Kong Instagram users' effort to position Hong Kong in the globalized social network space.
- *Intertextual references to slogans on protest sites*: Many of the Chinese hashtags had their origins from slogans seen on banners and card boards on the protest sites, such as #我要真普選 (I want universal suffrage) and #自己香港自己救 (Save our own Hong Kong by ourselves). Many of these express the supporters' political pursuits and needs.
- *Cultural references to Hong Kong (lyrics, slang, pop culture)*: A number of Cantonese hashtags point to locally produced internet memes and lyrics that were popularized during the movement. For example, #四點鐘許sir (four o'clock with superintendent Hui) refers to a public Facebook page devoted to a Hong Kong police officer who made public press releases at 4:00 p.m. daily during the movement. #我地大家 (together we . . .) comes from the lyrics of the Cantonese pop song 獅子山下 (Under the Lion Rock). It also points to an incident during the movement where a 28-meter banner with the slogan "I want universal suffrage" was hung on the Lion Rock by climbers.

These practices can be interpreted as a result of the complex interplay between englobalization and deglobalization of internet resources. *Englobalization,* according to Blommaert (2011) refers to global circulation of resources, while *deglobalization* occurs when globally circulated resources are deployed and reappropriated locally and given new local meanings and functions. Instagram, as a globalized social network, is taken up by Hongkongers as a site for showcasing locally produced resources. These include images of social events, which are annotated by hashtags written in local linguistic resources – Cantonese, traditional Chinese characters, and Chinese–English mixed code.

Using multilingual hashtags also serves as a powerful linguistic practice for authentication (Varis et al., 2011; Androutsopoulos, 2015). Here, *authentication* is understood as a process of people claiming authenticity, or realness, which can be realized discursively through semiotic resources including language and images (Karrebæk et al., 2015). Since the umbrella movement, the word *real*, or 真, has become a productive adjective in various pro-democracy slogans in Hong Kong, as in "I want real universal suffrage" or "真香港人" (real Hongkonger). The use of multilingual hashtags on Instagram can be understood as Instagrammers strategically making visible what Hongkongers speak, read, and write; what

Hongkongers should do with their linguistic resources; and, eventually, what makes a "real" Hongkonger.

Another example of Hongkongers defending their local linguistic resources is the "Kongish Daily" Facebook page (Wong et al., 2016). Kongish (a blend of *Hong Kong* and *English*) is a stylized variety of Hong Kong English which has been popularized on the internet and in various pop culture media in Hong Kong. It draws on translingual resources from English, Cantonese, Hong Kong Cantonese internet slang terms, transliteration, code-mixing, and being deliberately ungrammatical (e.g. *goodest* for *best*). Because Kongish Daily was launched soon after the umbrella movement and many of the posts are politically satirical, Kongish has been named a "language of protest" by the local newspaper *South China Morning Post* (SCMP, 2016).

LANGUAGE PLAY AND IDENTITIES IN SOCIAL MEDIA

Linguistic resources do not always serve immediate communicative and discourse functions in online writing. In Chapters 2 and 3, I discussed examples of how multilingual users deploy their resources of scripts and languages for the sake of playfulness and creativity. Language play online is not new and was found to be an important practice in early research on CMC. In IRC, for example, ASCII arts and playful screennames (e.g. cLoNehEAd, uh-uh) were found to be closely related to identity performance (see Bechar-Israeli, 1995; Danet, 2001). For multilingual web users, resources to play with or remix include languages, orthographies, and typographies. In multilingual digital writing, playing with words and combining scripts and typographic symbols from different languages allow users to perform and represent aspects of identities including being competent bilinguals (Jaworska, 2014), gender (Vaisman, 2011), being cosmopolitan citizens (Peuronen, 2011; Zhang, 2012), and even the construction of "otherness" (Leppänen and Häkkinen, 2012).

An immediate effect often presented by playful alternations of codes is *humor*. German speakers of English, as noted in Jaworska (2014), have creatively combined English and German to a comic effect in discussion forums. Often, these "digital code plays", as Jaworska puts it, deliberately violate expectations of code-switching in speech. In her data, Jaworska found that jokes in an online forum used by English-speaking Germans in Britain contained heavily anglicized German. An example is when the English verb *to pop* (as in "my ears are popping") is literally translated into German as *poppen*. Although this verb exists in German, it is indeed a taboo word in German that refers to sexual intercourse. Such form of "semantic play" was found to be quite prevalent in Jaworska's data.

Bilingual play involving English has also been identified among mainland Chinese internet users. In popular Chinese web spaces such as Weibo (a microblog service) and Douban (a social network site), all linguistic levels of Putonghua (Mandarin Chinese) have been twisted and played with, as reported on in Zhang (2012, 2015). Phonologically, in Chapter 3, we already saw how part of the word *holiday* was replaced by homophonous Chinese words. Morphologically, Chen (2007) found that in Taiwan, web users created new words through attaching English suffixes (such as the progressive *-ing*) to Chinese words, as in 期待ing (anticipating) and 招募ing (recruiting). Interestingly, such playful deployments of English and Chinese have been extended to official government web spaces. In December 2011, the police department of a Chinese city posted the wanted notice shown in Example 4.3 on their official Weibo site.

Example 4.3 Wanted notice on Weibo

#百团乐GO# "天清地宁"为答谢广大网民,现推出抓逃惊爆价！求秒杀！求给力！你还HOLD住吗？

清单如下：. . .

[personal details of fugitives listed here]

(Translation: #Hundred groups happy-go#: To thank the general netizens, "Clear Heaven and Peaceful Land" offers surprise fugitive hunting price! SecKill! Geili! Can you hold it? List of items: . . .

(Zhang, 2012: 45–46)

Rather than posting in a formal and official wanted notice style, the post in Example 4.3 adopts a hybrid and multimodal style that combines linguistic practices of a wanted notice, popular internet slang in China, as well as the writing style and page layout of Taobao, the biggest online shopping site in China. The discourse style of the notice strongly resembles a "group purchase" promotional event on Taobao. Adopting a seller–buyer discourse style, the ad begins with a short paragraph urging buyers (referring to the readers) to buy (meaning to catch) the fugitives to get an attractive sum of reward. It is then followed by the code-mixed expression 百团乐GO (hundred groups happy-go), in which *go* is homophonous with 购 (buy) in Chinese. 求秒杀 (sec-kill requested) and 求给力 (geili requested) are common Chinese internet expressions in that *sec-kill* (to kill in a second) is used on Taobao to urge buyers to act fast before something goes out of stock, and *geili* literally means "to give power". HOLD住 (holding) is another common code-mixed internet expression. In the announcement, the question 你还HOLD住吗? roughly means "Can you resist the temptation?" (i.e. the reward). This is then followed by a list with personal information and portraits of the fugitives. This exchange may be extremely complicated to readers who are not familiar with not only Chinese but also the particular discursive style of the Taobao shopping site. However, a sense of exclusive insider joke is exactly the effect that the writer of this notice intended to create.

In both the German and Chinese examples, the hybrid and playful deployments of linguistic resources on the internet, or digital code play, allow writers to display not only their linguistic competence and metalinguistic awareness, but also their multivocal identities (Zhang, 2015). The fact that language play has moved beyond the private sphere to official domains, as in the case of official microblogs in China, reveals the ways in which institutional domains, especially local governments, attempt to play down their identities as authorities, so as to establish rapport with the general public through vernacular and local linguistic practices.

There are also cases where multilingual language play does not involve actual uses of multiple languages. On YouTube, there are music videos of non-English-language songs that are given subtitles in English words that sound similar to the lyrics in the original language. These are often called *buffalaxed* videos because they had their origins from someone with the YouTube username "buffalax". Example 4.4 is the opening verse from the subtitles of the Tamil song "Kalluri Vaanil", retitled by buffalax as "Benny Lava".

Example 4.4 Excerpt from Benny Lava

My loony bun is fine Benny Lava!
Minor bun engine made Benny Lava!
Anybody need this sign? Benny Lava!
You need a bun to bite Benny Lava!
Have you been high today?
I see the nuns are gay!
My brother yelled to me
I love you inside Ed!

Drawing on the incomprehensible or misheard sounds of Tamil and lexical items in English, the resulting subtitled texts are incomprehensible to speakers of either English or Tamil. From the point of view of the subtitle writer, however, the subtitles are meant to be a joke. Although some commenters of the video take a critical stance and see buffalaxing as a possible behavior of racism, most commenters take it lightheartedly and find the videos hilarious. Linguistic competence is not entirely irrelevant here, as the subtitles are meant to be laughed at by those who know English only. However, the entertaining nature of the video is not solely created by linguistic resources; rather, it is a result of the visual (the music video), the sounds (the singing and the sounds of a foreign language), as well as the mistranslated subtitles that allow the West to playfully construct the identities of the Oriental others (Leppänen and Häkkinen, 2012). Leppänen and Häkkinen further argued that the practice of buffalaxing is comparable to inserting Asian languages in English fan fiction, in that both cases involve the use of the codes of "others" to construct the identities of people from other cultures. These practices give rise to new forms of translingual and translocal encounters across cultures.

NEGOTIATING IDENTITIES IN ONLINE TEACHING AND LEARNING SPACES

Another domain where identity performances in social media are of particular importance is education. In recent years, digital media have offered the affordances to support language teaching and learning, as well as a space for learners to reflect on their own language use both online and offline. For example, second language learners have extensively taken up learning opportunities provided by language learning websites, discussion forums, and mobile apps. The learning activities on these platforms may be deliberate, incidental, or even recreational (Chik, 2015).

In translocal networks, such as discussion groups and chatrooms for Chinese immigrants in the US, young people were found to represent and reconstruct their language learner and multilingual identities through careful deployment of multilingual resources. Lam (2009) examined in detail how a Chinese immigrant, Kaiyee, developed her Cantonese and English abilities through blogging and instant messaging with three online communities. These networks include a local network of Chinese immigrants where participants combine Mandarin and Cantonese, a gaming network where African American Vernacular English is used, and finally a transnational network where Kaiyee would mix Mandarin with Shanghainese when chatting with other IM participants. Through these concurrent networks with different linguistic and national groups, not only did Kaiyee gain new knowledge of Cantonese and English, she was also able to expand her existing second language learner identity by interacting with other learners (Lam, 2013).

In the case of Kaiyee, she participated in mostly public spaces, where there is basically no restriction on membership and members are generally free to express themselves in any language of their choice. Language choice and code-switching become an interesting topic when they are negotiated within a more locally "controlled" linguistic environment. Community-based online spaces such as Facebook groups have been adapted as "edusocial" platforms to facilitate offline face-to-face classroom teaching and learning (Pollara and Zhu, 2011; Reid, 2011). When classroom linguistic practices are mediated by social media, there often exist tensions between an official medium of instruction and the students' language practices in private digital writing.

In my research on Facebook groups created for my English linguistics courses in Hong Kong, both the teachers and the students have been found to adopt hybrid linguistic and discourse styles (Lee and Lien, 2011; Chau, 2014). In face-to-face lectures, these courses are instructed in English, and students are expected to use English in their academic work. However, when academic discussions move to Facebook, mixed-language writing is observed in both the students' and the teachers' posts. Chau (2014) noted that the occasional addition of Cantonese final particles helped students to construct a friendly persona and establish solidarity among not only their fellow students, but also between the students and the instructor and tutor. In addition, Lee and Lien (2011) reported on students' opinions about their language choices in the course groups; the students generally perceived the groups as English academic groups, but mixed-language writing was acceptable if there was a good reason, as a student revealed in the following interview excerpt:

> Sometimes it's more formal when i write in Facebook group as it is like an academic platform. But it also depends on the nature and formality of that message. If it is not related to academic field, I would rather switch to some chinglish as it looks more friendly and funny.

Not only do students need to negotiate their language choices in social media, but teachers are also particularly conscious of how they represent and position themselves through language in social media. In Lee (2014), I examined a case study in which a preservice teacher in Hong Kong, Tony, addressed his student audience and friends through strategically deploying his linguistic resources on two Facebook accounts. As a preservice English teacher, Tony connected with the students in his teaching practice school through a second Facebook account, where he called himself "Teaching Tony". In this Facebook account, his language choice and style immediately indexed his teacher identity. First, he wrote almost all his posts in English. He said in the interview that it would have been "inappropriate" to post in Chinese there because, as an English teacher, it was part of his profession to adopt a consistent "medium of instruction" in school and on Facebook. By doing so, he also hoped to encourage students to write to him in English as well. The two status updates in Examples 4.5 and 4.6 are indicative of his language preference in his teacher Facebook.

Example 4.5

Time to [be] back to school again! Good luck for your homework and the coming tests!

Example 4.6

Dear 4R students,

I have put copies of three sets of reading practice paper in my cabinet outside staff room. Please come and get it yourself if you need them.

Both posts were directed to his students, but they could also be read by his other Facebook friends, including his colleagues. From the content of the status updates on his teacher Facebook wall, it is clear that Tony was using Facebook not just as a social network site to stay in touch with his students, but also as a teaching and learning tool where he was able to make class-related announcements and share English learning resources with his students. By contrast, on his private Facebook wall, intended for his friends and family only, Cantonese or code-mixed posts predominated while almost no post was written entirely in English (see Example 4.7).

Example 4.7

收到了Jim Scrivener 2010的新作<<Teaching English Grammar>>,如果我TP之前就買左就好lah. . . .

(Translation: Just received Jim Scrivener's new work in 2010 Teaching English Grammar. I wish I had bought this before my teaching practice.)

Compared to the posts in his teacher Facebook account, Tony's posts on his personal Facebook wall are much more heteroglossic in linguistic resources. Example 4.7 shows a mixed language post with Cantonese and English writing. In the post, TP is an abbreviation for the teaching practice sessions in which he took part at that time. The Cantonese writing is represented in characters as well as in Romanized form. For example, *lah* is a Romanized spelling of the Cantonese discourse particle 喇.

Tony also reported that he had been using Cantonese–English code-switching in other forms of computer-mediated texts, especially IM, for about 10 years. According to Tony's technolinguistic biography (see the Appendix), also discussed in Lee (2014), mixed-code chat resembles everyday conversations with friends. It becomes clear that Tony's linguistic practices on Facebook are largely shaped by what he used to do in older CMC platforms such as IM. The shared practice of code-switching between Tony and his friends helped maintain their in-group identity as Hongkongers. Using different linguistic resources for different kinds of posts was clearly a salient practice in Tony's online impression management; that is, Cantonese and code-mixed utterances indexed his playful, down-to-earth personalities, while English, often used when writing about more serious subject matters on his teacher Facebook page, was used to assert his more official work identity. Tony's juggling between his different self-identities on two Facebook sites provides a telling case of his identity management when facing different audiences. What Tony was doing was extending his offline identities as both a student and a teacher, as well as his offline relationships with others. Switching between languages on two Facebook accounts allows him to maintain these personae and relationships.

DOING IDENTITY WORK WITH MULTILINGUAL RESOURCES ONLINE

Digital writing is about writing oneself into being (boyd, 2008). Identities online are highly textually mediated and are constructed socially and linguistically (Barton and Lee, 2013). In this chapter, I have shown how various forms of identities are constructed, performed, and reappropriated through multilingual resources and practices. These include projecting cosmopolitan and glocal identities on translocal platforms such as fan fiction and Flickr, self-authentication through multilingual hashtags on Instagram, translingual digital play to express new kinds of multilingual identities, and the negotiation of multilingual resources for student and teacher identities. These instances of identity work through multilingual resources also point to what is happening in the superdiverse digital space, in that multilingualism online is not just about which language dominates or linguistic competence. In fact, in the examples of identity performances discussed in this chapter, all multilingual resources serve equally important indexical and symbolic values. In the case of fan fiction, Asian languages in the characters' dialogues were used to project the users' multicultural and cosmopolitan identities. This is similar to the practice of having a Chinese–English bilingual screenname on Flickr, in which the different scripts in the screen name symbolize cosmopolitanism. Cantonese, although less used as a written language in offline contexts, is not less important than English in digital writing during a political movement. On Instagram, Cantonese written in traditional characters indexed "authentic" Hongkongers during the umbrella movement. In contrast, stylized English in German and Chinese digital media not only expressed humor, but also allowed participants to present themselves as new multilingual beings in the superdiverse world. For buffalaxed videos, it is exactly the "incompetence" in languages that gave rise to new translocal relations. Finally, code-switching in English-based academic Facebook groups conveyed a sense of friendliness and thus facilitated what would have been considered formal academic discussions.

These findings have important implications for research methods. Tony's case demonstrates various factors that shape language choice in social media. These include his previous linguistic practices online, his self-image, and how he wants others to see him. His attitude toward mixing Chinese and English online is a familiar one, as he was used to doing so in IM, which he used frequently when he was a teenager. These details would not have been revealed by studying texts alone. A technobiographic approach (see Chapter 8) pays closer attention to the situated nature of social media, tracing changes in language use at different stages of language users' technology-related lives. A focus on past experience reveals how older practices may shape newer ones. When sharing their life stories, participants also revealed their folk linguistic theories and language ideologies through metalinguistic comments. People's ideological representations of multilingual resources will be the focus of the next chapter.

5

REPRESENTATIONS OF MULTILINGUALISM ON THE INTERNET

Overview

- Representing multilingualism on the internet
- Multilingual categorization and management of online content
- Imagining multilingualism
- Metalinguistic discourses on the internet
- Talking about multilingualism in the online world

REPRESENTING MULTILINGUALISM ON THE INTERNET

This chapter is concerned with what and how multilingualism is "thematized" (Kelly-Holmes and Milani, 2013) by people who may or may not be multilinguals themselves. Up to this point, this book has focused largely on multilingual texts online as linguistic output (Crystal, 2011); that is, multilingualism is realized through language choice and code-switching *in* the texts produced by web users. This chapter turns to the *representations* of multilingualism on the internet. Speaking of media in a general sense, Holmes and Milani (2013: 1) offer a useful distinction between language *of* the media and language *in* the media. The former is concerned with observing and analyzing linguistic phenomena in CMC texts, on which the previous chapters have concentrated; the latter, which is the focus of this chapter, is related to: "(1) issues of how language is used to organise, categorise or target media; (2) how (different) languages are used as content within the media; and (3) what kinds of metalinguistic discourse underpinned by which language ideologies take place in the media". Likewise, representations of *multilingualism* take many forms on the internet. First, multiple languages have been used to organize media content, such as Wikipedia entries. Wikipedia's language policy and users' discussion of it on the Wikitalk page have been of interest to researchers in recent years (Ensslin, 2013; Deumert, 2014). Second, in terms of languages as *content*, we have come across numerous English-based websites that deliberately draw on multiple languages to perform a global persona; an example is the way Flickr greets its users in multiple languages (discussed later). Third, web users, regardless of their linguistic backgrounds and knowledge, often talk about languages when participating in social media. For example, Flickr users who do not speak English as a primary language often undermine their English abilities in public comments (Lee, 2012). These three ways of representing multilingualism online are further elaborated in the rest of the chapter. Drawing on examples from various social media, this chapter also discusses how talking about languages and multilingualism online creates new language learning opportunities.

MULTILINGUAL CATEGORIZATION AND MANAGEMENT OF ONLINE CONTENT

The first perspective from which multilingualism is represented online is the use of languages to manage and organize online contents. An official website for a regional government may make its content available in an official language only. For example, the website for the city of Xi'an in China is available in simplified Chinese only, whereas the main website for the central Chinese government has three language versions: simplified Chinese, traditional Chinese, and English. Globalized platforms such as Facebook and Flickr were first launched with English-only interfaces. As these sites get used by people who speak different languages, they have been translated into various major languages of the world over the years.

Among the most popular globalized online media, Wikipedia is one that explicitly markets itself as a multilingual platform. The first line on the About page of Wikipedia reads: "Wikipedia is a multilingual, web-based, free content encyclopedia". The order of the adjectives clearly indicates the site's priority for its multilingual contents. As of July 2016, Wikipedia is available in about 290 languages. On the surface, Wikipedia appears to be a site that truly embraces linguistic diversity. For example, more than 100 of its language editions contain over 10,000 articles. Many of the entries are independent articles that are available in certain languages only, suggesting that the articles are not simply translations from one language to another. For example, although the topic "multilingualism" is available in over 50 languages, the articles all vary in length, focus, and illustration. In the English edition, multilingualism has a much more refined subcategorization of contents and more comprehensive discussion of relevant theories in light of scholarly research written in English. The photographic illustrations of multilingualism on the English Wikipedia are also wide-ranging. By contrast, the Chinese Wikipedia entry for multilingualism is relatively short and has fewer subsections. There is almost no mention of scholarly theories of the concept. Most of the images that are used to illustrate multilingualism are photos of Chinese-speaking regions, such as Beijing, as well as China Towns from different parts of the world. In view of these, Wikipedia has created a virtual linguistic landscape where users' knowledge is valued as they create and publish contents based on their own languages and cultural backgrounds.

While multiple languages are used to manage contents on Wikipedia, languages also get managed by the site's language committee. This committee is responsible for reviewing new requests for language editions and coming up with final decisions as to whether the language in question deserves its own Wikipedia edition. On the Language Proposal Policy page on Meta-Wiki, a space for discussing Wikipedia projects, the language committee provides a list of requisites for a language to be eligible and makes it clear that "requirements *must* be met by requests before they can be approved" (Language proposal policy, Wikimedia, 2016). One crucial and often controversial criterion for the setting up of a new Wikipedia language edition is "number of speakers", as also stated in the policy:

> Number of speakers necessary
>
> There should be enough speakers to form a viable community and audience. Whether a particular language qualifies depends on discussion.
>
> (https://meta.wikimedia.org/wiki/Language_proposal_policy/en)

This description also attempts to market Wikipedia as a democratic community which listens to users' opinions and comments about a new language proposal, which are posted on their discussion pages. In looking through their list of rejected and eligible languages, however, exactly how the committee makes decisions is still unclear to general web users. For example, Ancient Greek, first proposed in 2008, has been rejected for the reason of lack of native speakers. However, some supporters argue that if other ancient languages have their Wikipedia pages, Ancient Greek should enjoy the same status, as one commenter complains:

> Not fair that Ancient Greek doesn't have a Wikipedia. A lot of people know this language and are willing to contribute. If this Wiki shouldn't be opened, why does the Anglo-Saxon and Gothic Wikipedias exist?
>
> (https://meta.wikimedia.org/wiki/Requests_for_new_languages/Wikipedia_Ancient_Greek_3)

Wikipedia's language policy has been problematized by new media scholars, from its potential Anglocentrism (Ensslin, 2013) to its underrepresentation of small languages and being largely "a European affair" (Deumert, 2014: 67; see also Chapter 2). Whether Wikipedia is a truly multilingual space remains questionable. From the descriptions about multilingualism and its multilingual policies, however, it seems to take a rather narrow, restrictive view of multilingualism that is largely defined by competence and number of native speakers. Its language policy is based primarily on the English version, although the site claims to welcome multilingual translations of it. The case of Wikipedia's multilingual policy is similar to that of the community translation pages on Facebook (see Chapter 7), on which translations of contents are heavily regulated by official policies. While many global websites claim to offer multilingual affordances, one must bear in mind that most of them were founded by and are still managed by English speakers. As a result, these sites offer yet another set of examples of what is called *imposed multilingualism* (Dor, 2004), where local languages are controlled and even manipulated by authorities.

IMAGINING MULTILINGUALISM

A second form of representation of multilingualism is the deliberate *display* of multilingual resources that do not serve any actual communicative or discourse function. On Wikipedia, multilingualism is often on display multimodally. Verbally, the word *multilingual* appears in most of the general definitions and statements about Wikipedia; visually, names of languages and their respective scripts are made visible in major parts of the site. The logo of Wikipedia, for example, is the first thing we see at the center of its homepage, and it appears on the navigation bar on all pages. As shown in Figure 5.1, the logo resembles a globe composed with jigsaw pieces, most of which contain the onset or first character of the word for Wikipedia in different languages. Other characters are a random selection of interesting symbols from different writing systems, such as the Greek omega (Ω) (Deumert, 2014). In other words, the omega symbol is thought to be representative of the Greek alphabet, thus creating a kind of "branding" or emblematic effect for Greek (Sebba, 2015). The missing pieces on the top of the globe are also suggestive of Wikipedia's constant effort in search of new language editions. However, a closer look at the globe reveals that dominant alphabets are foregrounded and positioned in the center of the globe, including Latin, Greek, and Cyrillic. Other languages such as Thai and Hindi are placed at the edge. This spatial arrangement of the scripts on the globe, as Ensslin (2013: 87) argues, still encodes "marginalisation and discrimination of less empowered, non-Western languages".

The logo has undergone many revisions since 2003, due to some misrepresentations of the scripts (as discussed in Ensslin, 2013). Among these errors was the Japanese *katakana*, the syllabary writing system based on sounds, on the globe. The Japanese characters in the logo remained as ワイ from 2003 to 2010, which was considered by some as "ambiguous and confusing" as the first character was incorrectly rendered (Talk page, Wikipedia). It was not until 2010 that the initial character was replaced with what appears in the present logo, ウイ, the first syllable of the word Wikipedia in Japanese (ウィキペディア).

The erroneous and randomness of this "multiscriptual" logo illustrates the mere emblematic function of scripts in the digital space (Androutsopoulos, 2013b). As of early 2016, as many as over 290 language editions are available on Wikipedia.

Figure 5.1 Logo of Wikipedia Image © Wikipedia [accessed July 2016 at https://en.wikipedia.org]

However, they cover only less than 10% of the world's languages. There is still a long way to go for Wikipedia to transform itself into a truly multilingual platform.

In a similar way, the photo-sharing site Flickr greets its users in a new language every time they click through their profile photos (Figure 5.2), from Chinese, Spanish, and Swahili, to smaller languages such as Mäori, an indigenous language in New Zealand spoken by just a few thousand people. This gesture has received mixed reviews; some users say they enjoy being greeted in a different language every time they log on to the website; others find the feature "annoying" (The Help Forum on Flickr, 2007). At the time of writing, the multilingual greetings have been extended to Flickr's Facebook page. Since 2015, a feature photo alongside the message "Hello, Hola, Bonjour, Buongiorno, Bom dia, Guten Tag, 你好 Flickr! Happy Monday [or any day of the week]! Photo by [user]" is posted on a regular basis on the Flickr Facebook site.

In Chapter 4, I showed the ways in which web users have taken up the symbolic and emblematic values of their multilingual resources in CMC (e.g. allowing users to assert cosmopolitan identities). The same idea may also be applicable to the way website designers, who do not necessarily know the languages, make multilingualism visible on their sites. The scripts and greetings on Wikipedia and

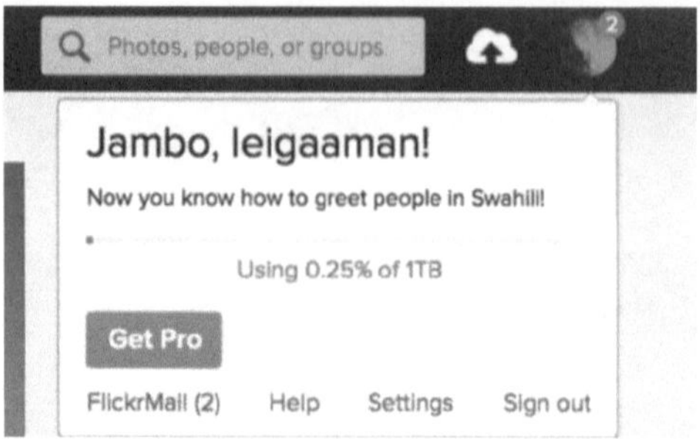

Figure 5.2 Flickr greeting a user in Swahili (Image © Flickr, accessed July 2016 at https://www.flickr.com/)

Flickr have enabled the site designers and owners to present the sites as "global" in superdiverse linguistic landscapes. To some, this may lead to what is called "fake multilingualism" (Kelly-Holmes, 2005; Lenihan, 2011; Thurlow, 2012), in that multilingualism on these often America-based English-speaking sites does not reflect authentic multilingual practices. While the logo and greetings are clearly marketing strategies, what seems to better describe the uptake of multilingual symbols, I argue, is that multilingualism is often imagined. In imagined multilingualism, authenticity is not always expected. As with word play in advertising, such simulated forms of scripts and languages are to be looked at, not read.

METALINGUISTIC DISCOURSES ON THE INTERNET

A third type of online representation of multilingualism is *metalinguistic discourse*, that is, how multilingualism is talked or written about on the internet. Metalinguistic discourses about multilingualism may be produced by anyone, regardless of their linguistic background; that is, one does not have to know many different languages in order to talk about multilingualism on the internet. It is human nature for people to be able to *reflect* upon language and their own use of it. Sociolinguists have long been interested in researching language users' attitudes toward linguistic varieties and accents through a range of field methods (Labov, 1984; Preston, 2004). One of the common methods for these studies is to elicit people's *metalanguage*, which comprises utterances that express beliefs and attitudes about language by nonlinguists. Before the digital age, discussions and discourses about language and its use tended to be limited to mass media, and ordinary people's participation was kept at a private level or could only be elicited through carefully designed research instruments. The internet and the affordances of social media provide unique opportunities for flows of public discussions about languages (Barton and Lee, 2013). Common spaces for people to engage in metalinguistic discussions about languages are the commenting spaces of various social media platforms.

The participatory web allows for the sharing of views and opinions about any topic openly and publicly. Metalinguistic comments are no longer confined to mass media or private conversations. Ordinary internet users from all over the world can engage in in-depth conversations, in any language, about their linguistic resources, their knowledge of them, and how these resources are deployed. All aspects of language have been the subjects of discussion on the internet. As detailed in Barton and Lee (2013: 110–115), a wide range of topics have been observed in metalinguistic discourses online: YouTube commenters have criticized "errors" of second language English grammar and spelling; features of Netspeak are thought to be harmful and affecting students' language standards; accuracy of translation has been discussed on Facebook; and people have made self-deprecating remarks about their own knowledge of English. Some of these have even become important practices of social networking among second language English users on Flickr (Lee, 2012). As more metalinguistic comments are generated, people are also becoming more aware of their own use of language, and thus languages have gradually become a common subject of online discussions. In what follows, I examine several types of discourse identified in online metalinguistic comments, including discourses of native speaker authority, prescriptivism, and self-reflection.

"Trust me hes perfectly fluent": asserting native speaker authority

Although it is essentially a video-sharing platform, YouTube is textually mediated and every video uploaded is surrounded by the written word (Benson, 2015). The commenting space on YouTube is undoubtedly a convenient conversational space for global users who speak different languages to come together to leave their opinions on any topic. Sometimes these comments are stimulated by the content of the video posted, but in many cases, the comments have little to do with the original content posted. Other comments are generated by the more "embodied" features such as facial expressions, gestures, and voices of the people in the videos (Lange, 2007). This then provides a space for talking about any unpredictable topics, including rather in-depth discussions about languages. In this section, I draw on the metalinguistic comments on a selection of YouTube videos featuring the performances of Wonho Chung, a stand-up comedian born in Saudi Arabia, to Korean and Vietnamese parents, and raised in Jordan. As discussed in Chun and Walters (2011), Chung's multicultural and multilingual background becomes an important influence on his performances, which are conducted mainly in Arabic but are often mixed with Korean and English. His humorous and parody videos toward Arab culture, coupled with his unique cultural identity have attracted many positive comments from viewers. One of the aims of Chun and Walters's study is to examine how YouTube serves as a site for negotiating and constructing stance and ideologies about race and identity. Example 5.1 is a selection of comments on Chung's performance in Dubai.

Example 5.1 Comments on Wonho Chung's stand-up comedy posted on YouTube

Commenter 1: HOLYYY SHITTTTTT. i was not expectin that at ALL he speaks better than my parents :|

Commenter 2: I love how he has an arabic accent when he speaks english!
Commenter 3: I like it................ sweet so so so coooooooooooooooooooooool....
الكوري الولد هدا مهضوم شو حبيتوا كتر كتير حلو جد ن ع (*Translation: Really, very very nice. I like him. He is amazing, this Korean guy.*)
Commenter 4: SPEAKS BETTER ARABIC THAN ME!!!!! for those of you who dont speak arabic, trust me hes perfectly fluent, no accent either

The four commenters in Example 5.1 all express knowledge of Arabic. Commenter 1 explicitly compares Chung's Arabic language skills to his or her own native knowledge of the language; commenter 2 is able to identify Chung's Arabic accent when he speaks English; and commenter 3 claims proficiency in Arabic through actually commenting in Arabic alongside the English comment. Commenter 4, while praising Chung for his fluent Arabic with "no accent", is self-deprecatory in that the commenter deliberately plays down his or her own native Arabic competence, by saying that Chung speaks better Arabic. In a way, the commenters position themselves as bilinguals who can speak or write Arabic while being able to make comments in English. A few other comments of a similar nature have been identified in the same thread. While praises for Wonho Chung's Arabic position Chung as a good or even better speaker of Arabic, Chun and Walters (2011) argue that these commenters take on a voice as *the* native speakers. By comparing Chung to their own native Arabic knowledge, the commenters also assert authority and authenticity of a native Arabic speaker that Chung lacks as an "outsider" with a strong Asian background. These comments can also be taken as acts of exclusion, implicitly marking Chung's otherness as a non-Arab.

In one of his performances, Chung switches between Arabic and English, and part of the English segment involves him humorously putting on a Filipino persona by mimicking a Filipino accent. While many of the commenters accept that as "funny" or "hilarious", some Filipinos are more cautious and attempt to debunk Chung's stereotyping of Filipino accent (see Example 5.2).

Example 5.2 "Not all Filipinos speak that way"

Commenter 9: the stand up comedy act in itself is funny but I hope he didn't made of fun of the FILIPINO accent..Not all filipinos speak that way1
Commenter 10: I'm Filipino and that's a close representation of the thick accent. Course we don't all sound like that because it really depends on how much we were educated in speaking English by our parents and schools.

Commenters 9 and 10 stress that not all Filipinos speak with Chung's accent, although both agree that he has done a good job as a comedian. YouTube has provided a space for Wonho Chung and comments from his audience (including YouTube comments) to collaboratively construct their stances toward Orientalism. Chun and Walters (2011) recommend that in examining the relation between Chung's performances and his viewers' feedback, it is also important to take into account the researchers' own "critical stance" regarding race (Chun and Walters, 2011: 269; see also Chapter 8). The researchers have the responsibility to acknowledge how their reflection and writing processes shape their understanding of the videos. This is because stances are always inferred. What is said (the form) may not be what the stance taker actually wants to convey (especially in the case of humor and irony). Such a contextualized, ideological, and critical perspective can be combined with insights from pragmatics which often aim to reveal not just what is said, but what is *not* said in utterances.

A similar phenomenon of constructing speakers' authenticity through metalanguage is also identified among German-speaking commenters on YouTube. Androutsopoulos (2013c) examines the ways in which German dialects are performed and negotiated in a corpus of videos tagged with German dialect names such as *Bairisch* (Bavarian) or *Berlinerisch* (Berlin city dialect). The two focal videos in the study are "Rinjehaun – Berlinerisch fur Anfanger" (Berlinerisch for beginners), which is a three-minute dialect lesson, and "MacBookAir auf Berlinerisch" (MacBookAir in Berlinerisch). Although the two videos deliver different kinds of content, the majority of commenters orient their comments to the general features of German dialects or how accurately they are delivered by speakers in the videos. Example 5.3 provides two of the comments on the MacBookAir video.

***Example* 5.3 Comments on "MacBookAir auf Berlinerisch"**

na ick find dit nicht so jut jemacht, dit kommt nicht so orjinal rüber, ick finds eher auswendig jelernt und uff gesacht

(Translation: Well, I don't find it that well done; it doesn't come across as original. To me it feels rather learned by heart and delivered.)

Boah ne, sorry, dit is aber ma so ja nich knorke! Da_ hatt er zwar die Vokabeln jepaukt, aber . . . dit klingt so ja nich Original-Berlinerisch . . . keene Stimmmelodie drin, weißte Keule!

(Translation: Oh well, sorry, but this isn't my thing at all! He did learn his lessons, but . . . it doesn't sound like original Berlinerisch, no vocal melody in there, isn't it mate!)

(Androutsopoulos, 2013c: 64)

Both commenters doubt the originality of the German dialect style used in the video. They also frame their comments through the process of "othering", using phrases such as "to me" or "this isn't my thing". Like Chung's commenters, they claim to possess more original or authentic knowledge of the dialect in question.

"The right accent": prescriptivism and the "purist" discourse on amateur language teaching videos

Video-sharing sites such as YouTube have also given rise to a set of new vernacular and multilingual practices. Studies have found that social media have served as important platforms for critical discussions, language learning, and cultural exchanges (Benson and Chan, 2011; Barton and Lee, 2013). For example, Benson and Chan (2011) analyzed a corpus of comments made on three English-subtitled (or fansubbed) versions of the Beijing Olympics theme song. In addition to the presence of multiple languages, their data exhibit significant evidence of language teaching and learning discourses among the commenters. YouTube also hosts abundant language lessons conducted by amateurs and nonprofessional teachers. In what follows, I discuss two YouTube videos, each featuring a "non-native" speaker teaching a second or foreign language. I focus on the language ideological discussions in the comments section and the discourses of prescriptivism towards non-native speakers of a certain language.

A Canadian YouTuber, "CarlosDouh", whose real name is Carlos Vidal, became a celebrity after posting his first video in his Learning Cantonese with CarlosDouh series; in each video he introduces a new Cantonese expression delivered in both English and Cantonese. Instead of teaching traditional lessons such as greetings or asking directions, he teaches one colloquial expression at a time, with the hope that this approach would attract more viewers. In a newspaper interview, he said: "Instead of teaching regular, everyday words which wouldn't get that many views, I started doing slang words and people just thought it was really funny" (SCMP, 2011).

The most popular video on Carlos's channel is called "I am a Hong Kong girl with 公主病 (Gung Jyuh Behng)", which has over 1.4 million views and over 1,000 comments as of January 2016. 公主病 literally translates as "princess sickness", which is a Cantonese slang word to describe women who are or want to be spoiled like princesses. In Example 5.4, the commenters engage in a conversation by drawing on multilingual resources from Korean, Cantonese, Japanese, and English.

Example 5.4 A multilingual language lesson among commenters

B: 공주병 Wow it sounds really similar haha

C: how does it pronounce?

B: Gong ju byung. Something like that.

D: That's cause Chinese is the oldest language compared to Japanese and Korean. The Japanese have 'kanji' and the Koreans have 'hanja', where they have taken some Chinese characters (called 'hánzi) and write them the same but pronounce them differently. They also have similar sounds for meanings, such as 공주병 and 公主病.

E: i totally agree with you~~~many Japanese and Korean vocabularies sound like Chinese. I found lots of Korean vocab have similar sounds to Hakka, one of the dialects of Chnese.

(Benson, 2015: 91)

The commenters here orient to pronunciation issues. However, rather than focusing on Carlos's pronunciation, these commenters have started their own mini language lesson by comparing what 公主病 may "sound like" in different languages. What dominates the thread of comments on this video is the commenters' lay linguistic theories and their discourses of prescriptivism (Jones et al., 2011). That is, to them, languages, in this case accents, are evaluated in terms of good and bad, accurate and wrong. This is illustrated in the thread in Example 5.5.

***Example* 5.5 "The right accent of Cantonese"**

Commenter E: Your pronunciation is perfect! . . .
Commenter F: WOW. LOVE UR ACCENT, YOU'RE REALLY SPEAKING THE RIGHT ACCENT OF CANTONESE! LOVE YOU!<3
Commenter G: Omggggg . . . dude, your accent is spot on . . . It's hilarious cuz I've never seen a white guy speak so good canto ! Good job man ! ;)
Commenter H: WOW! You got some awesome chinese cantonese language. And with no accent in it. Awesome. Others just speak it with an accent.

Like Chung's commenters, Carlos's viewers position themselves as the more authoritative speakers of Cantonese, who are qualified to evaluate Carlos's Cantonese as "perfect" and "awesome". That the commenters praise Carlos's accent as "the right accent" and "spot on" indexes their purist or prescriptive discourse (Thomas, 1991). To them, there is only one ideal way of speaking Cantonese, and that includes not speaking it with an English accent ("with no accent in it. Awesome").

On YouTube, a non-native speaker's language production, especially their accent, can easily become a subject of high praise, but the vlogger can at the same time be a victim of flaming (Lange, 2007). Another video that also features the teaching of the vlogger's non-native language, by contrast, is not as well-received as the one by Carlos. In 2010, a secondary school girl from Hong Kong, Ruby, posted a vlog on YouTube called "funny teacher saying foul language lols", in which she criticizes the spoken English of an invigilator in a public examination. The original video on Ruby's channel was taken down but has been reposted by other YouTube users. A range of multilingual practices, including code-switching, are noted in Ruby's original post, as well as her viewers' comments (Benson, 2010). In the video, Ruby first introduces the context of her story in Cantonese – that the invigilator spoke "primary school level" English at a public examination and she could not stand it. Then, she switches to English in the video because, as she explains, she wants the whole world to know some "common English errors" among Hong Kong people. According to Ruby, the teacher mispronounced *ask* (as *ass*), *zip* (as *sit*), *during* (as *diu2*, a swear word in Cantonese), and *sheet* (as *shit*). Ruby, while attempting to correct the teacher, comments repeatedly in the video that the teacher's pronunciation was "really really funny" and urges that all teachers should have a "higher English level". The video itself is only one site of metalanguage on this YouTube page. What is more interesting is that the video immediately became a YouTube sensation among local Hong Kong people when it was posted in 2010. What really interested viewers was not the issue of the bad English used by the school teacher; it was Ruby's English pronunciation and grammar in the video that immediately became the subject of a series of hostile comments. Many viewers criticized Ruby for her poor English (some targeted at her accent) and said that she should have taken care of her own English before criticizing others (see Example 5.6).

Example 5.6 "She speaks Chinglish"

WR: . . . She speaks Chinglish, so why is she making fun of others' Chinglish?
MML: I feel so sorry for your poor English. =) And is that Cantonese? she has very big accent for Cantonese . . . So Bullshit..LOL
KF: . . . honestly, dont pretend you know English cuz girl your chinese accent is for sure greater than your teachers . . .

Once again, these three comments demonstrate the familiar discourse of linguistic purism (Thomas, 1991) and the classic language ideology of "English with an accent" as a problem (Lippi-Green, 1997). The fact that Ruby's English pronunciation contains traces of Cantonese accent is considered by these commenters as laughable "Chinglish" or even "poor English". The comments also reflect the way in which Hong Kong people align themselves to exonormative models of English (Luk, 1998). This means that there is a strong preference for American or British English pronunciation rather than for local norms of "Hong Kong English". In Hong Kong, English with a Cantonese accent has long been socially stigmatized. Varieties adopted by native speakers of English are still considered to be more accurate and acceptable than those spoken with local Hong Kong features.

"I won!": multilingualism as added value

Another set of discourses identified in online metalinguistic discussions conveys a sense of "the more languages the better" attitude. In the context of minority languages movements, Jaffe (2007) observed a shift from the discourse of "bilingualism as deficit" to that of "bilingualism as added value". This shift is also evident in an online forum discussion of the status of Irish (Kelly-Holmes, 2013b) and in Wonho Chung's video commentary discussed earlier. As the conversation develops, Chung's viewers shift their discussion to what appears to be a competition of multilingual competence among several commenters, as shown in Example 5.7.

Example 5.7 Number of languages known

Commenter 5: wow i wonder how many languages this guy knows. geeeez I'm really amazed.wallah I know some americans that have trouble speaking their own language. lol Luv
Commenter 6: I'm an 18 year old american and I know 4 languages :p
Commenter 7: WOW really which languages? Are you fluent in all of them? I personally know english spanish and arabic but I'm not fluent by any means. Luv Jenn-
Commenter 6: English Spanish and Italian . . . currently learning Korean. There's others that I want to learn though. I love languages and cultures.
Commenter 8: 15 year old,I know 2 languages,learning third and next month starting to learn two more.. I won! haha
Commenter 6: This isn't a contest hun. I was just making a point that I don't fit the "dumb america" stereotype.

The thread is initiated by commenter 5, who is impressed with Chung's multilingual competence and wonders how many languages he knows. She also makes a remark about how "some Americans" cannot even speak their own language, presumably English here. Commenter 6, who appears to be American, immediately distances himself from commenter 5's stereotype of Americans, by listing the four languages he can speak. Another comment follows and claims to have outdone commenter 6 by boasting that she will know five languages. This game ends by commenter 6's announcement of "This isn't a contest". These comments clearly reflect the commodification of multilingualism in the global world, that knowledge of many languages is seen as asset and "added value" (Jaffe, 2007; Heller, 2010; Kelly-Holmes, 2013a).

The global internet has enabled multicultural and multilingual interaction to take place among users from around the world. Thus, to web users, being able to master different languages would automatically mean being able to reach to a wider and more global audience. As a result of multilingualism as added value, a discourse of monolingualism as deficit also emerges in online discussions (Kelly-Holmes, 2013a). This is often constructed through what is called self-deprecating metalanguage, that is, utterances where language users downplay their own linguistic abilities (see Barton and Lee, 2013: 114–115). On a number of new media platforms, participants who claim to know only a language other than English are often apologetic for not being proficient in English. In Black's (2009) fan fiction research, a young fan fiction writer, Nanako, plays down her own English writing skills when seeking help from others. In her opening and closing author's notes, she writes: "Important note: English is my second language and I only spoken it for 2.5 years. So please excuse my grammar and spelling mistakes." She also refers to her "bad writing" and states "I am not a good writer". Similar remarks were also observed in the comments and personal profiles of participants on the photo-sharing site Flickr, as reported in Lee (2012):

> Well, i'm sorry, my english is so shitty ugh because i'm french so sorry.
>
> PS Sorry, my English is as lousy as my photography, it's not my first language

While these self-evaluations suggest that knowledge of English is still an essential linguistic commodity with high communicative value on the web, the 'monolingualism as deficit' discourse coexists with these self-deprecating comments about one's own English:

> I am embarrassed that I only speak English, . . . while I expect everyone else to read and write English.
>
> I'm Sorry, English is My Only Language.

In these comments, the writers are embarrassed or even feel sorry for being able to speak only English. They consider that being monolingual or not knowing additional languages limits their participation on a globalized site like Flickr. These comments also lend support to the changing perceptions from the internet being a largely Anglophone space to a more heteroglossic one. All in all, undermining one's own linguistic knowledge is not simply an act of modesty. These self-deprecating comments serve multiple social functions on the internet. In the case of fan fiction, identifying herself as weak in English, Nanako in fact creates a supportive environment for her fan fiction writing where she can elicit constructive feedback from other writers. Similarly, being apologetic about being monolingual in English on Flickr immediately indexes one's willingness and readiness to participate in a global community. Talking about their limited linguistic knowledge also becomes an important self-positioning act where people articulate their language ideologies, create supportive environments, and exercise authority.

Social media also provide people with supportive spaces to share ideas that they would not have shared in public or in face-to-face communication. When people reflect on being a multilingual, they often situate their discourses in their lived experiences and express how being a multilingual, or not being one, has had positive or negative impact on their lives. Academics are among the many professions whose lives have changed drastically because of digital media (Barton, 2015b). On July 19, 2015, Caroline Magennis (who is @DrMagennis on Twitter), a lecturer in literature from the University of Salford, UK, initiated a conversation by posting the following question on Twitter (Times Higher Education, 2015):

> What are the challenges of being an academic from a less privileged background? Questions of "fitting in" but also practical issues?

What appeared to be an academic debate gradually developed into a discussion about languages. Many of the tweets foreground the importance of being a polylingual academic, as shown in the following tweets by academics from different disciplines.

User 1:	@DrMagennis Access to languages is a big one in history. If you learnt Latin at school/had £ to learn mod langs abroad, life much easier.
User 2:	@user1 @DrMagennis also sense that lack of languages is result of personal failing, rather than result of less privileged education
User 3:	@DrMagennis the exhaustion of simultaneously having to learn 2 new languages, the language of academia and the one of the middle-class.
User 4:	@DrMagennis @user2 Internet dictionaries with audio files are a huge step forward!
DrMagennis:	@user1 @user2 as a PhD student I was eternally concerned with pronouncing critics names wrong that I'd only known written down
User 5:	@user3 @DrMagennis Public speaking and debating were not part of my schooling hence anxiety about pronunciation and my accent!

On the one hand, although these academics are not linguists, these comments all draw attention to languages and how multilingual knowledge is considered more powerful and crucial in academia. What is also noteworthy about these tweets is the ways in which academics have taken up the affordances of Twitter to create a supportive environment where they feel at ease to share their concerns with like-minded colleagues. On the other hand, the commenters resort to a narrow and traditional understanding of being a multilingual – that being a multilingual means being a polyglot (Kramsch, 2015). As many examples in this book illustrate, multilingual practices need not involve the coexistence of multiple languages. It is also interesting to note that they situated multilingualism in the context of their lived experiences; to some, this reminded them of their schooling, their encounters with other academics, their academic advisors, and how knowing more languages could facilitate their learning of academic language, such as pronouncing names of foreign authors. Others point to the old sociolinguistic question of accent and social class, and that academics may be disadvantaged or "less privileged" by having an accent or not adopting a middle class linguistic style. Once again, the familiar discourse of linguistic capital (Bourdieu, 1990) is prevalent in this thread of tweets by academics. For these academics, language, or multilingualism, to be precise, is definitely treated as profit (Heller and Duchêne, 2012) that not only allows them to progress in their profession, but also serves as an indicator of academic identity.

"You got fat": engaging in multilingual activity in one language

In Chapter 1, I pointed out that a multilingual activity can be carried out without involving multiple languages and that one does not have to be proficient in many languages to engage in a multilingual activity. Multilingual practices are concerned with the *process* of working with different languages rather than just the product of it. Davies (2007) cited a relevant example from Flickr. An American Flickr user, "Saffron," once posted a photo of herself wearing what she referred to in the photo caption as a "throw" given to her by her grandmother. A Norwegian Flickr user, "Astrid", then came across this photo page and asked Saffron what the word *throw* meant in the photo description. Saffron explained:

> Hi astrid, yep like a small blanket i guess . . . a pashmina is also used as a wrap and a chunky scarf. this is about a metre and a half long knitted wrap. i guess my grandmother would call this a shawl. I would think a poncho refers to material with a hole in the centre so you slide it on rather than wrap it. aaah such confusion.
>
> (Davies, 2007)

Apparently, this comment is written in English only. From the view of learning, this interaction is packed with discourses of language learning and can be seen as a "mini-English lesson" (Benson and Chik, 2010: 68). In other words, multilingual practices can be understood as the ways in which groups and communities of people experience and do things that involve more than one language. Web users with different linguistic and cultural backgrounds discussing languages online has become an increasingly common practice, especially on sites that facilitate public commenting, such as Flickr and YouTube. These are just some of the new practices that people never found themselves doing before the age of digital media.

A less-examined but equally important digital writing space for metalinguistic discussions is news commenting. Today, traditional print media have a strong online presence where they provide more interactive spaces to engage readers. Most mainstream newspapers have their own websites and share popular stories on Facebook, Twitter, and Instagram. All of these platforms enable public commenting. Not only does news commenting allow people to respond to a certain news story, it also gives rise to social interaction between the news writer and the commenters (Diakopoulos and Naaman, 2011). In July 2015, an article with the title "Cantonese 101: 'You got fat!' means 'Hello' in Hong Kong" was published in *South China Morning Post*, the major English-language newspaper in Hong Kong. In the article, the author, Mary Hui, a Hongkonger who had returned from her first year study in the US, was disturbed by the fact that she was always greeted with a comment about her being fat or fatter. She then also gives examples of Cantonese expressions which mean fat in a positive way such as *fung mun* (plump or busty). She ends the article by saying that she is "in doubt as to whether people mean I'm *fung mun*, or just fat, fat, fat". This was then followed by a debate among her readers about their views on Cantonese greetings with connotative meanings (Example 5.8).

Example 5.8 Readers' comments on Cantonese greetings

ML: I left HK in the 60. When I went back to visit later on, I found myself out of style if I said "have you eaten rice yet?" I've learn to say, "You good?"

KCC: My dad once told me that the greeting, "Have you eaten?" came from the time when Mao ruled China and everyone did not have anything to eat. That eventually became their greeting.

P.S. I'm a Chinese living in the Philippines.

WBJ: Sorry to say this, they say it not only because of just greeting, HK style greeting is "You are fatter!" OR "You are skinnier!" OR "You are more muscular! OR "You are prettier!", and they mean it. It happens like that because HK people quite focus on other's appearance, and they dare to say it in front of your face.

DBG: You can't literally translate this into English, there's always a cultural significance in such greetings, this does not literally say you've become fatter, but rather that you're eating well and you are healthy, not malnourished like in the old times.

DM: In Hong Kong it means "You're doing well". Stop imposing Western "victim" whining on local culture.

SC: To answer your question, Mary, they think you're fat, fat, fat. The only cultural aspect to this story is that people here will say it to your face.

In terms of language choice, the first interesting point to note in these comments and, in fact, in the original article, is that there is no trace of Cantonese writing. The whole communication is delivered in English only, even though the subject matter of both the original article and the comments is Cantonese. All of the target Cantonese expressions are given literal translations such as "You got fat" and "Have you eaten rice?" instead of the actual Cantonese expressions 肥咗喎! and 食咗飯未? The language choice here is a situated one – it is the immediate context of the newspaper that shapes the commenters' language choice; the *South China Morning Post* attracts mostly English readers including not only bilingual Hongkongers but also a large proportion of readers who only know English such as expatriates and tourists. Yet, this monolingual English interaction, like the throw example on Flickr, can still be seen as an instance of people engaging in multilingual practices in the digital world. They is evidence of people doing things with and about two or more linguistic resources. Looking more closely at the comments, ML and KCC both make reference to their personal experiences with Cantonese greetings; the other commenters, however, find the author's understanding of the Cantonese expressions problematic and attempt to offer alternative explanations. DBG, DM, and SC all point to the fact that a cultural explanation is missing in the article. In so doing, these online commenters collaboratively contribute to and even enrich the news story in question. This example clearly illustrates what Bruns and Highfield (2015) refer to as *collaborative news curation*. That is, digital media have provided new affordances for ordinary web users to actively participate in a new form of journalistic experience, including sharing and evaluating news stories publicly. They also offer new opportunities for people to co-construct and even reconstruct news production practices, such as correcting factual information in a news story.

TALKING ABOUT MULTILINGUALISM IN THE ONLINE WORLD

This chapter has moved from exploring multilingualism as linguistic output to multilingualism as *metalinguistic discourse*. It considers the ways in which multilingualism is represented, or thematized, by both producers and consumers of web contents. Various types of multilingual representation have been explored in the chapter. First, from the angle of web designers, multilingualism is adopted by globalized sites, such as Wikipedia, to organize and manage web contents. However, the multilingual policies imposed by these sites are often prone to public criticism and debates, as illustrated by user debates found on Wikipedia's Talk pages. Second, globalized platforms that claim to embrace multilingualism often project or imagine multilingualism as just a marketing tool or an emblem. The logo design of Wikipedia is an example of the emblematic use of multiple scripts so as to brand itself as a global space in which anyone and everyone can participate. For some digital media companies, displaying symbols of multilingualism is even a profit-making strategy. Contrary to traditional arguments about multilingualism being devalued (e.g. Blackledge, 2005), the examples in this chapter have shown that multilingualism is a powerful resource and practice for many globalized websites. From displaying scripts on Wikipedia's globe to multilingual greetings on Flickr's homepages, languages are treated as valuable sources because they help international organizations project a global image and expand their user bases (Jaworski, 2015). However, as Kelly-Holmes and Milani (2013) note, these websites are caught up in a "multilingual deadlock". That is, on one hand, it is encouraging to see a large number of languages being represented on global sites such as Wikipedia; on the other hand, their multilingual strategies and policies are always prone to criticism because they will always exclude some languages.

Much of the discussion of the chapter has been devoted to various kinds of online metalinguistic discourses *about* different languages. Drawing on examples from user comments on YouTube, Flickr, and online news, I have identified several recurring discourses about multilingualism. These include assertions of native speaker authority, linguistic prescriptivism and purity, multilingualism as added value, and the ways in which people talk about multilingualism in one language. To ordinary users of the web, metalinguistic discourses about multilingual practices serve multiple social functions. Talking about multilingualism allows people to construct themselves as different kinds of language users in different multilingual encounters: Sometimes they may project themselves as native speakers or more proficient speakers than other people; in different contexts, they may be second language learners and less proficient speakers. Various examples have shown that self-deprecating discourses about one's knowledge of languages can help widen participation on the web and create a supportive environment for multilingual practices to develop. Eventually, these metalinguistic discussions also contribute to the development of linguistic diversity on the internet. The level of people's attentiveness to multilingualism on the internet is unprecedented. In a way, this is a manifestation of a basic property of human language – reflexivity. Reflecting on language use is key to the development of language and its impact on people's everyday lives.

6

MINORITY LANGUAGES AND THE INTERNET

Overview

- The scope of minority languages in CMC
- Minority languages on the internet: Opportunities and challenges
- Representing lesser-written languages on the internet
- The future of minority languages online

This chapter zooms in on languages that are commonly recognized as "minority languages" and the extent to which they are present on the internet. In Chapter 2, I briefly discussed the problem of minority languages being underrepresented on the web, such as the limited coverage of isiXhosa on Wikipedia. At the same time, the internet has provided unprecedented opportunities for speakers to make minority languages more visible to the world (Cunliffe and Herring, 2005; Danet and Herring, 2007). In addition to exploring classic examples of minority languages, this chapter is particularly interested in those spoken varieties that have no standard writing systems, such as Luxembourgish and Egyptian Arabic. On the internet, these spoken forms are written in creative ways, enabling users to assert their local identities (see Chapter 4). Linguistic varieties to be discussed in this chapter include Catalan, Welsh, Irish, and Chinese dialects. Some of these are considered successfully represented online, while others are under threat with the rise of the internet. Examples to be discussed in this chapter illustrate how speakers of these languages represent their local languages in their digital writing. Before I move on to consider particular languages, some fundamental questions need to be addressed. What does it mean to be a minority language? How "minor" are they? And how has the concept been used in the context of computer-mediated communication? This chapter then examines both the opportunities and challenges that digital media present to minority languages. As it becomes clearer in my discussion in the rest of the chapter, the internet can present both opportunities and challenges to minority languages, depending on a complex combination of factors.

THE SCOPE OF MINORITY LANGUAGES IN CMC

The topic of minority languages has always been a complex and political one. One of the first questions has always been what term should be used and what *minority* actually means (Cunliffe and Herring, 2005). The word *minority* in *minority language* has been interpreted in different ways and it indexes different meanings in different domains in society. In linguistics, researchers of minority languages often cite definitions by government organizations and census reports. A commonly cited definition is one stated in the Charter for Regional or Minority Languages of the Council of Europe, which defines minority languages as languages that are:

i. traditionally used within a given territory of a State by nationals of that State who form a group numerically smaller than the rest of the State's population; and
ii. different from the official language(s) of that State.

(European Charter for Regional or Minority Languages, 1992: 2)

From this point of view, minority languages refer to those lesser used varieties with smaller numbers of users than the official languages (as also adopted by the European Bureau for Lesser Used Languages). Others have used the term *minority languages* as an equivalent to "heritage" (e.g. Lee, 2006), "threatened" (e.g. Fishman, 2001), "indigenous" (e.g. Dyson et al., 2007), or "endangered" languages (Gorter, 2008). In reality, however, these different terms have very distinct meanings and cannot be used interchangeably. For example, an indigenous language is not necessarily an endangered one. For some languages, their minority status is not a straightforward issue. Some of these languages may even have an official status but are seen as threatened for a complex combination of reasons. Irish, for example, does not fall neatly into any of the above usages. Lenihan (2013: 25) draws on a range of official and academic sources when capturing the complex situation of Irish:

> The Irish language is the first official language of the Republic of Ireland as per article eight of the Constitution of Ireland (1937). . . . In addition, Irish is an official language of the EU since 2007. . . . Despite all these status provisions, the Irish

> language is classified as "definitely endangered" on the UNESCO (2009) vitality scale. 1.77 million of the 4.5 million resident Republic of Ireland population claim to be able to speak Irish, but 1.16 million of these either report they never speak the language, or speak it less frequently than weekly (Central Statistics Office, 2012).... Kelly-Holmes (2011: 512) terms it a "privileged, minoritised language...."

In the context of CMC, research into minority languages online is growing yet limited. There has been no consensus as to what minority should refer to in internet research. Often, the term minority languages in CMC research has been used "as a convenient label, rather than as an attempt to adhere to a particular school of thought" (Cunliffe and Herring, 2005: 133). In the multilingual CMC literature, minority languages may refer to one of the following broad notions:

- lesser used/spoken varieties (e.g. Cunliffe and Harries, 2005; Lenihan, 2011, 2013)
- endangered languages with few living speakers (e.g. Thomsen, 2002; Rau and Yang, 2009)
- local dialects or indigenous languages (e.g. Sperlich, 2005; Liu, 2011, 2012)
- lesser-written languages (e.g. Warschauer et al., 2007; Themistocleous, 2012).

The first three meanings are more common and self-explanatory; the fourth category deserves some elaboration. By "lesser written", I refer to languages or varieties of a language that are typically spoken, or those that can be written but have no standard writing systems, or those that used to be written but the written forms are no longer in use today. Cantonese, for example, is not normally considered to be a minority language because of its large number of speakers worldwide. However, it is essentially a spoken language and written Cantonese is still an emergent variety that has no standardized system of writing. It is thus of particular interest to linguists and CMC scholars as to the extent to which such a largely spoken variety gets written in CMC environments. For example, a characteristic that is frequently reported in studies of lesser-written languages online is Romanization, that is, when users consciously spell words in a spoken language by borrowing sounds from English or other Roman scripts. While it is not my intention to adopt a particular definition of minority language, I use the term mainly within the scope of these four meanings.

MINORITY LANGUAGES ON THE INTERNET: OPPORTUNITIES AND CHALLENGES

There have been numerous efforts and initiatives which have used the internet to maintain or revitalize minority languages through language archiving and documentation (Eisenlohr, 2004; Galla, 2009). One way of approaching this is where technology and multimedia tools have been developed to facilitate the learning of endangered languages. One of the earlier studies of such efforts is Warschauer's (1998) ethnographic study of how Hawaiian universities make use of networked technologies such as bulletin boards, to preserve old Hawaiian materials, including audio- and print-based ones. These online materials and archives have also facilitated the learning of Hawaiian language through multimedia means.

Another approach is where projects aim to document minority languages. A notable example is the Endangered Languages Project powered by Google (www.endangeredlanguages.com). When Google launched the project in 2012, it promised to "put technology at the service of the organizations and individuals working to confront the language endangerment by documenting, preserving and teaching them" (Endangered Languages Project website). Partners and collaborators in this project include experts and institutes in the area of language preservation. The project website includes multimedia resources that document samples of different endangered languages including audios, videos, and texts. The project is well received by mass media reports, claiming that the project, covering 3,000 endangered languages, would eventually bring hope to these languages facing extinction. Ironically, the interface of the website is only available in a few major languages including English, Chinese, French, German, and Spanish. Questions remain as to whether such a project can truly facilitate language revitalization and what linguistic diversity actually means to the project administrators.

To date, empirical research on the actual impact of technologies on endangered and minority languages is limited. Scholarly discussions of the relationship between minority languages and the internet are twofold. As the following quotes suggest, some researchers focus on how smaller languages have been marginalized online because of the dominance of English and some larger languages, while others speak of the global internet as providing new possibilities for minority languages:

> Computers and the Internet present new opportunities [e.g. digital archiving and e-learing] for the preservation and transmission of endangered languages and cultures, both inside the language community and around the world.
>
> (Rau and Yang, 2009: 207)

> If the new medium is already heavily influenced by a particular language (i.e. English), it should come as no surprise that the new medium will further enhance the status of that language at the expense of an endangered language. . . . if this "natural" evolution drives us ever closer to global monolingualism and monoculturalism, we will be doomed to become what some of us fear most: clones of each other.
>
> (Sperlich, 2005: 76)

These opposing views provide further evidence for the complex relationship between minority languages and the internet; they also reveal that the global internet may present different consequences for minority languages in different situations. Rau and Yang's (2009) project starts from a language documentation and e-learning point of view. The aim is to develop an e-learning environment for Yami, an endangered Austronesian language of Taiwan, with the hope to revitalize the language and its culture. Certainly, the internet does provide the technological possibilities for that. Nevertheless, whether the users take up such opportunities is another question. That also explains the skepticism in Sperlich's (2005) quote. What Sperlich is more concerned with are the actual users' behaviors and linguistic attitudes toward maintaining their own native languages. The fact that many globalized websites and interfaces are still based on English may encourage and reinforce users' preferences for English whenever they participate in online activity. However, the fear that this shift will eventually result in "monolingualism and monoculturalism" may be an exaggeration. At least, studies in multilingual CMC have shown that users of minority languages have identified creative ways to communicate in their languages online, as will be seen in a later section.

Language shift

Arguments similar to Sperlich's (2005) have been put forward by a group of CMC researchers who believe that the internet further threatens endangered or minority languages. These discussions are closely tied to the idea of digital divide between a dominant language, which is usually English, and a minority language. This division may lead to at least two problems, one of which is people's shift from their local languages to a dominant language such as English; another problem is the explicit exclusion of minority languages on globalized sites. One obvious example is that the main interface language on many international websites or the official sites of global companies is still English, despite the coexistence of a few other "major" languages, such as French and Spanish. Over the years, scholars have reported on cases of explicit language exclusion online. For example, Irish was banned by America Online in its chat group "Peace in Ireland" (Ostler, 1999). A similar form of exclusion has been noted on LEONENET, a mailing list devoted to political events in Sierra Leone (Wright, 1996). Krio, although being the country's lingua franca, has given way to English on LEONENET. Posting messages entirely in Krio is considered by some of the list participants as "impolite", and English is used as a more "neutral" language. While supporters of English feel that the use of Krio in the list excludes non–Sierra Leoneans, Krio speakers think the language can facilitate communication among Sierra Leoneans in the diaspora. In this case, the list participants' sense of audience clearly shapes their attitudes toward their local languages online.

Abandoning indigenous languages for English is especially evident among younger generations of internet users. In a questionnaire survey of school children's actual use of and attitudes toward Irish, Fleming and Debski (2007) found that school children in Ireland rarely use Irish when they are online, and many are not even aware of the existence of Irish language websites. Most of them reported using only English both in their offline and online writing. Even if Irish is used, it is limited to certain situations such as "speaking in secret when there were non-Irish speakers present" (98). Irish speakers' attitudes toward the language are also discursively constructed in online forums. In a thread about political leaders holding a debate in Irish on the site boards.ie, participants reveal the "truth" about the situation of Irish – that few people can actually speak and understand the language, despite compulsory education. English is still seen by many forum participants as the key language of Ireland (Kelly-Holmes, 2013b). At the same time, the forum gives rise to a new discourse that views bilingualism in Irish and English as "added value" (Jaffe, 2007; Kelly-Holmes, 2013b).

Language shift is also noted in bilingual Welsh–English web environments. Although Welsh is an officially recognized language of Wales and is spoken in other parts of the world, such as Argentina and the US, it is still one of the six minority languages in the UK according to the European Charter for Regional or Minority Languages. The language is under threat for various reasons, including massive migration, decreasing number of speakers in the rural area, and an influx of English monolingual users. Cunliffe and Harries (2005) analyze the linguistic and functional patterns of Pen i Ben, a bilingual Welsh–English web community for Head Teachers in Wales. Their findings reveal a mixture of linguistic patterns among the regular contributors on Pen i Ben. While some contributors use exclusively Welsh, others write posts only in English for professional purposes. For those who use both English and Welsh, Welsh serves mostly social purposes, suggesting its perceived minority status in formal contexts. Although these findings may be suggestive of the successful creation of a bilingual online community, language shift to English is observed over time. In general, there is a significant decrease in Welsh use and an increase in English postings. Cunliffe and Harries recommend social and technical solutions for both majority and minority languages to coexist in a web environment. Socially, web hosts may be introduced to moderate topics as well as the languages used on the site; technically, machine translation can help overcome the language barriers on bilingual forums to a certain extent.

Climent et al.'s (2003, 2007) discussion of the situation of Catalan is also concerned with machine translation (see Chapter 7). Although Catalan has enjoyed official status in the province of Catalonia, it remains a minority language within Spain. In a corpus of over 500 postings on a Catalan university's Usenet, Climent et al. have found that over 70% of the posts are written in Catalan. However, when replying spontaneously to non-Catalan postings, there is a tendency for the Catalan-speaking participants to shift to Spanish, the dominant language. Climent et al. have proposed that one way to maintain the use of Catalan is to develop a sophisticated machine translation system for the language such that speakers would feel free to use their own language on the internet. The topic of translation will be further discussed in Chapter 7.

In a similar way, the Niuean language, the language of the Niue Island in the South Pacific Ocean, is considered to be threatened because of the introduction of networked technology (Sperlich, 2005). At first glance, the internet seems to have provided new opportunities for the preservation of Niuean. With only 20,000 speakers worldwide, the language has its own domain name (.nu) and the speakers on the island have reasonable access to the internet. As with other minority languages, websites and forums in Niuean allow the Niuean diaspora to communicate with each other in their own local language. However, the tendency to shift to English is also observed among Niuean speakers on the internet. In Sperlich's analysis of English and Niuean use on OKA-KOA, which is the first Niuean discussion forum and message board, 67% of the messages are written in English only, while as few as 5% of the posts are only in Niuean. The rest of the posts are mixed English and Niuean messages. In addition, the study shows that the more technical the language is in the message, the more likely it is for the writer to switch to English or mixed codes. Sperlich argues that the shift to English should not be entirely dependent upon the content of the message. To him, many cases are due to people's "lack of effort or competency" (Sperlich, 2005: 75).

All of the above studies point to one conclusion – that is, although the internet has provided the necessary technology and affordances for documenting minority languages, it has not stopped speakers of minority languages from shifting to a more globally circulated language, which tends to be English. Although studies as such are still limited, these examples all lead to the speculation that endangered languages will become more endangered with the rise of the internet. In view of this, there have been calls for reversing language shift through various means, such as language revitalization projects, development of machine translation, and online minority language documentation and teaching (Climent et al., 2007; Hogan-Brun, 2011). Reversing language shift, however, is not a straightforward task and does not rely solely on technological development; likewise, language shift is not necessarily linked directly to new technological advancement or the internet. As several authors have argued, the success of language revitalization online depends largely on the language users. Language users' choices are closely tied to their attitudes, ideologies, perceptions, motivations, and efforts (Eisenlohr, 2004; Sperlich, 2005; Fleming and Debski, 2007). On the contrary, there have been successful cases where minority language users have come up with solutions to represent their languages in online interactions. The rest of the chapter focuses on the ways in which minority language speakers write in, or write about, minority languages on the internet.

Minority languages in digital communication

The discussion so far has suggested that "big" languages are still privileged on the internet, even by speakers of minority languages, and this trend may eventually lead to the possible decline or even death of minority languages. Nonetheless, there are ample examples of how minority languages have been given new lives by their speakers on the global internet. The fact that CMC relies heavily on written language has provided new opportunities for smaller languages to regain their status in written form (Paolillo, 2007). One of the earlier successful cases is Assyrian, which is traditionally spoken by a small minority group in various parts of the Middle East, including Iraq, Iran, Turkey, and Syria, but does not have a static nation state. McClure's (2001) discussion of spoken and written representations of the use of Assyrian among Assyrians in America reveals significant presence of Assyrian on the internet. Instances of Assyrian–English code-switching are evident in chatrooms, newsgroups, and other kinds of online publications. Contrary to studies that see code-switching between English and a minority language as a tendency to shift (e.g. Sperlich, 2005), McClure considers this as an important opportunity to maintain minority languages through written means. This case also supports the idea that the success of language maintenance on the internet relies on its speakers. Assyrian is visible online not because of advanced internet technology. In fact, because Assyrian does not have a standardized alphabetical system, typing in Assyrian is not a straightforward matter. Nonetheless, the Assyrian web users have not abandoned their native language because of technical constraints. Rather, they have managed to transliterate or "spell out" Assyrian using the Roman alphabet. In English-based chatrooms and newsgroups, Assyrian is often used to discuss topics related to their ethnicity. This suggests that writing in Assyrian not only allows participants to actively revive their local language, but it is also a clear act of self-positioning, so as to strengthen their Assyrian identities.

The technological affordances of social media sites such as Facebook and YouTube have played a significant role in revitalizing minority languages and regional dialects. The representation of German dialects on YouTube discussed in Chapter 5 is an excellent example of how self-generated comments in social media give rise to public awareness of linguistic issues surrounding their local dialects (Androutsopoulos, 2013c, 2015). As noted in the previous section, Welsh is found to be marginalized in earlier studies of CMC such as forums and emails (Cunliffe and Harries, 2005). However, more recent studies have reported extensive use of Welsh on Facebook (Honeycutt and Cunliffe, 2010; Morris et al., 2012).

In an exploratory examination of the use of Welsh on Facebook, Honeycutt and Cunliffe (2010) present evidence that the Welsh language is actively present on Facebook, both in groups and people's personal profiles. Among the 368 groups they have analyzed, 236 of them contain Welsh, and 76 of these have Welsh-*only* content and descriptions. And of the 137 profiles examined, a significant proportion (67.3%) contain Welsh. Morris et al. (2012) further support Honeycutt and Cunliffe's findings about the presence of Welsh on Facebook. Focusing on only young Welsh speakers, Morris et al. draw on both quantitative and qualitative data in investigating their use of Welsh in social network sites. Their online survey results reveal that over 30% of the respondents use Welsh on Facebook, either alone or alongside English, and that older pupils would use more Welsh than the younger ones. The study also discovers regional variation in Welsh use on Facebook – that is, the young people from the northwest of Wales, where Welsh is spoken in school and other domains, tend to use more Welsh on Facebook than those from the southeast, where Welsh is not used widely. This once again suggests that the maintenance of minority languages relies on users' effort in both online and offline settings. Regardless, compared to earlier research that reports almost no use of Welsh on the internet, both studies indicate that social media sites such as Facebook are likely to play a significant role in maintaining Welsh and other minority and endangered languages. One reason for minority languages to be more visible in social media is that "networks of strong ties may help speakers resist pressures towards language shift" (Honeycutt and Cunliffe, 2010: 226). The fact that Web 2.0 media encourage user-generated contents has empowered users to normalize their use of Welsh on Facebook as a bottom-up effort, rather than following explicit language policies imposed by official bodies. However, Lenihan (2012, 2013) and Hendus (2015) have problematized Facebook's explicit top-down language policy, which can limit users' language choice. This will be further discussed in Chapter 7.

REPRESENTING LESSER-WRITTEN LANGUAGES ON THE INTERNET

In addition to making visible some recognized minority languages, what is particularly interesting about digital media is that they have given new lives to what I call *lesser-written languages*. At the beginning of the chapter, I defined lesser-written languages as those that are never or rarely written in offline contexts, although many of these are not given an official minority status. In the following, I will look at three examples of such languages: Cantonese (and other Chinese dialects), Luxembourgish, and Egyptian Arabic.

Cantonese and other Chinese dialects

Cantonese is the largest variety of the Yue dialects. It is used widely in southern China, especially in Guangdong Province, Hong Kong, and Macau. It does not have an officially recognized writing system, and few written materials are available for the teaching and learning of the language variety. In order to write in Cantonese, a common method is to borrow the standard script of Mandarin Chinese. Cantonese characters that do not have equivalents in the standard written Chinese script have to be invented by the users. While texts in Cantonese are increasingly popular in domains such as advertising and sometimes even mainstream Chinese newspapers, the status of written Cantonese remains controversial (Li, 2000; Snow, 2004). It has never been given any official status. With the increasing use of Mandarin, or Putonghua ("the common speech") as the official spoken language in official domains and as the medium of instruction in schools in Hong Kong and other Cantonese-speaking regions, the future of Cantonese is threatened. This has also made processing Cantonese on the computer difficult and challenging. At the same time, a number of studies of CMC have reported significant use of Cantonese writing, at least in informal online communication such as ICQ instant messaging and casual email messages (James, 2001; Lee, 2007b). James (2001) examines Hong Kong undergraduate students' use of Cantonese sentence final particles (SFPs) in their ICQ messages. SFPs in Cantonese, like intonation in English, are mostly added to the end of an utterance to indicate pragmatic functions such as assertion and contradiction. In James's analysis of a small sample of ICQ exchanges between two of his students, the SFPs are all inserted at the end of English utterances, as shown in the ICQ messages by Philip in Example 6.1.

Example 6.1 ICQ messages with Cantonese final particles

Philip: work hard *ar*. u will have present on friday *wor*. and have 4 midterms. don't day dream *la* i will support you mentally *ar*

(James, 2001: 12)

There are altogether four SFPs in Example 6.1. The first three particles, *ar*, *wor*, and *la*, serve the functions of softening the speaker's tone, reminder, and persuasion respectively. The fourth *ar* is similar to the first *ar*, which expresses a friendlier tone. What is noteworthy here is, first of all, the spelling of these words is largely self-generated by the writer, possibly through learning from other ICQ users' writing conventions online; second, an English sentence ending with a Cantonese SFP is rarely heard in face-to-face spoken discourse, except when the speaker is deliberately playful. James points out that while such a form of writing may be seen as bad English to some in formal contexts, the insertion of Cantonese SFPs into their English messages allows the student participants to create "solidarity and collectivity" (James, 2001: 15).

A similar observation has been made in my own study of ICQ exchanges among Hong Kong young people (Lee, 2007b). In Lee (2007b), not only do the participants Romanize Cantonese particles, they also draw on a wide range of playful and effective methods to enable themselves to write in Cantonese in online interaction, such as morpheme-by-morpheme literal translation (as described in Chapter 2). In addition to multiple forms of representation, the participants use Cantonese for specific purposes as well as to particular target audiences. For example, they frequently add particles to their English messages when chatting with their Cantonese-speaking friends only, but would never do so when chatting with Mandarin speakers. Such careful selection of resources according to the linguistic backgrounds of chat partners reveals that the participants are constantly aware of their own and their chat partners' linguistic identities. In other words, adding Cantonese final particles when chatting with Cantonese-speaking chat partners is a practice which they draw upon to highlight their Cantonese identity (see Chapter 4). The use of Romanized Cantonese in CMC has also been observed among overseas Cantonese-speaking students in Britain and the US (Lam, 2004; Fung and Carter, 2007). For these student communities, using Romanized Cantonese is more than just being playful or creative (Fung and Carter, 2007); having shared norms of written Cantonese enables these non-native English speakers to create a collective ethnic identity in diasporic online platforms. While earlier studies of Cantonese CMC focus on online chatting, written Cantonese continues to be a prominent feature in social networks frequented by Hong Kong people, including Facebook (Lee, 2011) and mobile instant messaging applications such as WhatsApp (Hafner et al., 2015). Warschauer et al. (2007: 304) suggest that the internet "fosters written communication in dialects and language that previously were used principally for oral communication". The presence of spoken Cantonese online has provided strong support for this claim.

The future of written Cantonese lies in its speakers. Although today the language is still alive on the internet and in various popular domains, the promotion of Putonghua as the Chinese national lingual franca, as well as the government's push for Putonghua literacy education in Cantonese-speaking regions, is likely to put written Cantonese at risk of diminishing (SCMP, 2014; see discussion in Gao, 2012). Online forums and social media have certainly offered the space for generating discussions about promoting and preserving Cantonese as a regional lingua franca. They at least provide the opportunities for speakers to write Cantonese through characters or Romanization. There have also been initiatives from academics in Hong Kong to promote Cantonese via social media. An example is the HKU Cantonese Facebook page (www.facebook.com/hkucantonese) maintained by a group of Cantonese linguists at the University of Hong Kong. The site offers Cantonese learning resources as well as information, news, and discussion about Cantonese. On YouTube, numerous videos can be found with free Cantonese lessons from basic greetings to idiomatic expressions. The teachers of these videos range from amateur native speakers of Cantonese, to professional language teachers, and sometimes even learners of Cantonese. What is also encouraging is that Cantonese has its own Wikipedia edition with over 46,000 articles, as of July 2016. Instead of calling the language edition 粵語 (Yue/Cantonese speech), which is suggestive of it being spoken, the Chinese name for the Cantonese Wikipedia edition is 粵文 (written Yue/Cantonese), in which 文 literally means "inscription". Wikipedia provides ample discussion spaces (e.g. the Talk pages) for users to share their thoughts about this regional variety of the Chinese language. For example, the controversial issue of the naming of the Cantonese edition has been a key topic in the discussion pages. The success of preserving written Cantonese does not only rely on producing texts in Cantonese. Ongoing discussions *about* the language can also enhance public's awareness of it. Sometimes these discussions are triggered by major offline events. For example, Gao (2012) examines metadiscourses in online forums defending the status of Cantonese after the 2010 protecting Cantonese movement in Guangzhou, China.

Apart from Cantonese, there has been limited research documenting the presence of other Chinese dialects on the internet (Liu, 2011, 2012; You, 2011). Speakers of the Chinese dialect Shanghai Wu on the internet, like Cantonese speakers in Hong Kong, have experienced the problem of what is called 有音無字 (sounds without characters) in Chinese (Liu, 2011). Shanghai Wu, or what is more commonly referred to as Shanghainese in English, is a Chinese dialect spoken widely in Shanghai and surrounding regions in China. On the SHN website (www.shanghaining.com), a forum devoted to *Shanghaining* (the people of Shanghai), participants have invented ways to represent their dialect for their online communication because there is no standard writing system for it. Unlike Cantonese internet users who Romanize Cantonese, Shanghainese speakers represent their dialect mostly in characters, through the process of phonetic borrowing of similar sounding characters in the standard Mandarin character set.

Liu (2011) notes that this kind of online dialect writing serves multiple functions on the site. First of all, the deliberate display of Shanghai Wu expressions on the homepage is emblematic (see Chapter 3 and Chapter 4). That is, rather than serving any real semantic or pragmatic function, Liu (2011) argues that the Shanghai Wu characters in the logo and other buttons vividly display or visualize the sounds of Shanghai Wu. Second, the lack of a standard writing system also implies the lack of constraints in deploying linguistic resources, thus giving rise to linguistic creativity on the site. For example, in the BBS column of the "Overseas Shanghainese" section on the SHN website, the negative word /və/ in Shanghai Wu (equivalent to the Mandarin *bu* 不 or *not* in English) is written in a wide range of characters, such as 伐, 弗, 佛, and 拂. At the same time, for frequently occurring words in Shanghai Wu, participants of the site have developed their own conventions and norms to represent them. For example, 謝謝 (thank you) is always written as 下下 on the site. The creative use of Shanghai Wu on the forum has facilitated the maintenance of Shanghai Wu and its associated culture to a large extent. With the internet becoming an indispensable part of people's everyday lived experiences, not only is it possible for speakers to maintain dialects in their spoken and written forms, but they have also been able to take up new opportunities to collaboratively create and negotiate a writing system for their own native speech. At the same time, it can be seen as an act of people's rebellion against the regularization of Putonghua (or Mandarin) as a common language in China (Liu, 2011).

Luxembourgish

Another lesser-written language that has been revived on the internet is Luxembourgish. As with Cantonese in Hong Kong, Luxembourgish is mostly used as a first spoken language in Luxembourg. The other official languages, French and German, are used in official written domains. But unlike Cantonese, Luxembourgish used to have a standardized written form in the 19th and the 20th centuries. However, it is not taught within the education system, where French and German are the common languages of instruction. With only about 400,000 speakers worldwide, Luxembourgish is classified by UNESCO (2013) as a "vulnerable" language, meaning that it is only used in restricted domains. In addition, given the introduction of English as a compulsory subject and the growing number of migrants from neighboring countries (France, Belgium, and Germany), Luxembourg is increasingly becoming a superdiverse and multilingual country (Belling and de Bres, 2014).

The linguistic diversity of Luxembourg is also reflected in digital media. Crucially, written Luxembourgish has regained vitality in digital communication. In a longitudinal analysis of the language practices found in the Luxembourg-based Facebook group Free Your Stuff Luxembourg, Belling and de Bres (2014) found significant use of Luxembourgish among members over time. The group, which aims to allow the citizens of Luxembourg to exchange goods for free, was created by a Romanian migrant. The main languages of communication of the group changed with the change of major events and language policies of the group. When the group was first launched, most members used English as a lingua franca. This is partly due to the fact that the group administrator posted mostly in English and that the group descriptions were all in English only. However, the group started to become more multilingual after the introduction of a multilingual policy by the group administrator as a result of a debate among members about languages used in the group. The newly added policy reads:

> About the language on this group:
>
> Because Luxembourg is a small but multilingual country, not everybody will be able to understand each other's language in this group. Therefore we do ask you to write in those languages that you master best. If there would still be any problem of

> understanding, people should be able to kindly ask for a translation to the person who does GIVE or NEED something or to the admins or to anybody else in this group who can help people understand each other. Thank you!
>
> (Free Your Stuff Luxembourg group, Facebook, 2015)

Since then, other languages, including Luxembourgish, emerged alongside English, and the group became more heterogeneous linguistically. Toward the end of the study, when there was no administrator regulating language use of the group, interestingly, Luxembourgish almost became the sole language of communication in the group. This finding, on the one hand, further supports the use of social media as a main site for language revitalization. On the other hand, members' resistance to multilingualism and the decision of using exclusively Luxembourgish pose new questions for current understanding of what superdiversity means in the digital world (Belling and de Bres, 2014).

As with many other lesser-written languages, Luxembourgish is made visible through metalinguistic discourses online. Also focusing on Luxembourgish, Horner and Krummes (2011) present a case study of the ways in which Luxemburgish is represented on YouTube. Numerous Luxemburgish language lessons can be found on YouTube. Horner and Krummes's study looks into the written comments made on some of these language lessons. A major topic that commenters discuss is the status and categorization of Luxembourgish as a language. In doing so, the commenters employ a number of strategies: They code-switch between Luxembourgish and English, and they articulate their historically rooted linguistic ideologies. What is worth noting is that these ideologies are sometimes conflicted, due to the geographically and culturally diverse background of the commenters. For example, commenters cannot come to a consensus as to whether Luxembourgish should be categorized as a German dialect; others engage in a debate as to whether the language actually sounds more like French or German. All these debates and arguments among commenters, although at times hostile, can in turn positively reinforce the status of Luxembourgish as a language. The fact that these language-related discussions are posted on a global platform for public viewing is unprecedented. Public commenting enables speakers of Luxembourgish to discuss their native language in a way that they would have never experienced outside the internet. The growing use of the language online also raises their awareness of their own language use, thus contributing to the revitalization of both spoken and written Luxembourgish.

Egyptian Arabic

Egyptian Arabic is perhaps one of the most successful cases of a lesser-written language being made prominent online. In Egypt, classical Arabic is the language having high prestige and is widely used in print-based texts including books and newspapers, as well as in formal speech contexts. Colloquial Arabic, by contrast, is used in informal spoken situations only. Processing Arabic on the computer is never straightforward. This is partly due to the lack of software standards and people's unfamiliarity with Arabic typing, as they first learn computing through English. These in turn encourage the adoption of a Romanized form of Egyptian Arabic writing (Warschauer et al., 2007). As with speakers of Romanized Cantonese and Luxembourgish, speakers of colloquial Arabic borrow sounds from English such that they can "spell" Arabic words in their online writing. According to Warschauer et al. (2007) discussed in Chapter 2, although English is still a dominant language in the formal emails of young professionals in Egypt, there is no sign of complete language shift to English yet. On the contrary, colloquial Arabic, which used to have very limited use in offline contexts, is used extensively in informal email and online chatting in Warchauer et al.'s data. While a large proportion of Egyptian Arabic is used alongside English in code-mixed messages, a few are written entirely in Egyptian Arabic. The motivation to use Romanized Arabic online is partly driven by users' willingness to preserve their indigenous culture and assert their identities online. The participants in the study also stress that their use of English online is not an attempt to abandon their Egyptian identities. Studies conducted later than this one have also observed sustained use of Romanized Arabic in social media such as blogs and Facebook. Some even argue that the popularization of Romanized Arabic online gives rise to a new e-Arabic register (Bjørnsson, 2010; Daoudi, 2011). For example, Daoudi (2011) has noted that Arabic itself is changing at all linguistic levels due to the vast use of digital communication. Features of what she calls *e-Arabic* are found at both literal and figurative levels on the internet. One of the characteristics of e-Arabic is to replace an Arabic character with a number that is similar in shape. Daoudi (2011) argues that the spread of shared e-Arabic norms makes Arabic more accessible especially to those who are not educated in modern standard Arabic.

There are many other successful cases where minority, endangered, or regional dialects are used extensively on the internet, which I cannot discuss in detail within the scope of this book. These include the presence of written norms of isiXhosa in SMS (Deumert and Masinyana, 2008), which was discussed in Chapter 2; the representation of various Jewish languages, such as Yiddish and Ladino (Judeo-Spanish), as reported in a special issue edited by Benor (2011); and the use of Pittsburghese, a regional dialect of American English, in an online discussion board (Johnstone and Baumgardt, 2004).

THE FUTURE OF MINORITY LANGUAGES ONLINE

The foregoing discussion on minority languages on the internet has spelt out some ongoing issues revolving around the circulation and representation of languages in an increasingly superdiverse world. Taking a broad definition of minority languages, the studies reviewed in this chapter suggest a highly complex relationship between minority languages and the internet. Both challenges and opportunities have been observed. Despite efforts to improve multilingual technologies online, such as machine translation, language documentation, and multilingual versions of websites, many minority languages are still left out or underrepresented on global sites. In some cases, even with the necessary technological affordances, people are reluctant to use their native languages online for various reasons, leading to a shift to English or other dominant languages that people perceive as the lingua franca of the internet.

Despite these unresolved issues, there are many successful cases of a minority language being used extensively on the internet, such as the cases of Assyrian, Welsh, Cantonese, Luxembourgish, and Egyptian Arabic presented in this chapter. Speakers of these different minority or lesser-written languages have taken up the affordances and available resources in new media to represent their own speech. For speakers of languages that do not have a standardized script or that do not use the Roman alphabet, a common method they have adopted is to Romanize their languages by borrowing sounds and the alphabet from the English writing system. Although this may not be a satisfactory solution, it can at least overcome some of the language barriers minority language speakers may have experienced and facilitate communication among diasporic communities worldwide. More importantly, the internet provides an important platform for what used to be minority or lesser-written languages to become public and more visible in mainstream domains. Studies have also observed the close correlation between linguistic creativity and the representation of minority languages, as in the cases of e-Arabic and Romanized Cantonese.

There is a consensus within the limited body of literature on minority languages in digital media that Web 2.0 social media have played a central role in maintaining and revitalizing endangered and vulnerable languages (Jones and Uribe-Jongbloed, 2012). Several authors have discussed the potentials of Facebook for the teaching and learning of minority languages. Lessons of endangered languages on YouTube help document and preserve the languages in question, and the public exposure of such languages is also unprecedented. What is perhaps the most distinctive feature, as discussed in Chapter 5, is the affordance of public commenting available on most social network sites. Engaging in conversations and debates about their native tongues allows users to generate new knowledge for speakers of the languages and for the general public, who may have never heard of the languages in question. Thus, the pedagogic implications of social media for minority languages are not to be overlooked. On YouTube, fan-translated or fansubbed videos are definitely another important way of raising public's awareness of minority languages or smaller dialects. The relationship between translation and the internet will be further discussed in Chapter 7.

The future of minority languages is not solely dependent upon the medium or how advanced the technology is. Whether the internet presents opportunities or threatens minority languages depends on a complex combination of factors. The impact of the internet on minority languages varies from context to context. The continuation of such linguistic debates is fruitful in that they can certainly inform our understanding of linguistic diversity in the digital world. Future studies need to take some of the following issues into consideration when exploring the status of minority languages on the internet:

- *Language policies online and offline*: On the one hand, whether or not a language is used online is a political issue; it is often triggered by the tension between a dominant or official language policy and a lesser-used language.

On the other hand, while official language policy of a region plays a role in shaping its people's ideologies and attitudes, it is often the existence of an internal language policy of a website that affects its users' language choices (Lenihan, 2013; Belling and de Bres, 2014). Some have even recommended imposing a minority-language-only policy on the web so as to encourage the use of local languages online (e.g. Fernandez, 2001, cited in Cunliffe and Harries, 2005). However, as Cunliffe and Harries (2005) argue, such a policy ironically encourages a separation between languages. Linguistic marginalization or exclusion is exactly what advocates of minority languages are *not* in favor of.

- *Speakers' attitudes and willingness to use their languages online*: "[A] successful online community requires media producers and consumers" (Cunliffe, 2007: 134). This is also true of a successful representation of a language online. The future of a language is up to its speakers, including their perceptions of the status of the minority language concerned (Davies, 2004; Sperlich, 2005; Climent et al., 2007). As some of the examples in this chapter have already demonstrated, speakers of minority languages are ready to shift to larger languages. Not everyone sees the urgency of preserving languages or the value of human languages in the same way.
- *Technological possibilities and constraints for the minority languages concerned:* The extent to which a minority language gets represented also relies on the technological affordances available for speakers of that language, that is, how easy or difficult it is to produce texts in that language on the computer. These affordances include inputting systems, machine translation software and applications, and the availability of domain names in a particular language. This eventually affects the amount of content produced and the level of access to content in the language concerned.
- *Context of use*: The findings from most of the successful cases of minority language use indicate that minority languages are mostly present in interpersonal and informal contexts, such as Welsh in personal profiles on Facebook and Cantonese in private instant messaging. These suggest that minority language use is still limited to a certain extent.

7

ONLINE TRANSLATION AS A MULTILINGUAL PRACTICE

Overview

- Translation and the internet
- Community translation as a vernacular multilingual practice
- Getting things done through translation practices in the mobile world
- The future of online translation practices

If one is to name a multilingual activity that has been drastically transformed by new media and technologies, it would have to be *translation*. The changing affordances of machine translation and computer-assisted human translation have not just made translation between languages faster and more affordable, but they also present new forms of communication and linguistic practices. Seeing translation as both a social and a multilingual practice, this chapter is not so much about showcasing the translation tools or online dictionaries available in the market, nor is it an overview of the development of machine translation; rather, it is more concerned with how ordinary people take up the affordances of translation tools to facilitate their multilingual practices and engage with international audiences in their online participation. It also discusses the increasingly popular practice of *community translation* that takes place in various social media, such as fansubbing on YouTube and using the translations application on Facebook.

Finally, the chapter explores how practices of online translation impact on people's everyday lives in the mobile world.

TRANSLATION AND THE INTERNET

Machine translation and computer-aided translation began in pre-web times. Before the age of free online translators, handheld electronic dictionaries were invented to provide instant machine translation and pronunciations of user-input sentences. In the early days of personal computing, basic tools such as spell-checkers and grammar-checkers on word processors provided people's earliest experiences with computer-assisted translation technology in its broadest sense. Free translation tools and multilingual dictionaries on the web, such as SYSTRAN's Alta Vista Babelfish and the Merriam-Webster online dictionary, began to emerge in the 1990s. This was also a time when English still dominated the web, despite the fact that many users did not use English as their primary language outside the internet (see Chapter 2). In view of this, websites of governments and official organizations started to become available in multiple language versions. University websites around the world, for example, are increasingly bilingual or even multilingual. Universities from regions that do not use English as an official language are likely to include English as a "first additional language" for international site visitors, on top of the national language version which is meant for local readers (Callahan and Herring, 2012).

Advances in free online translators have presented new translingual practices and even changed the way people engage in multilingual activities in society. One of the best known free online translators is Google Translate (http://translate.google.com). First launched in 2006, Google Translate supports machine translation of over 90 languages. It not only offers translation of text-based contents, but also translates speech, images, sounds, and even videos from one language into another. The Google Chrome browser can be configured to automatically translate webpages written in a foreign language into the browser's default language. According to its original creator, Franz Josef Och (2005), Google Translate does not adopt a grammar rule analysis in their translation method, but a statistical machine translation method. This means that instead of teaching the machine basic grammatical rules such as word order and sentence structure of the target language, the translation results are based on the frequency of word collocations, that is, sequences of words that appear together in a sentence. This also means that the program produces a translation that is more frequently occurring in its corpora; this is believed to be a more accurate and reliable method of machine translation. For the past 20 years, there has been growing interest in applying corpus linguistics in machine translation (Dajun and Yun, 2015).

While being as reliable as it can, statistical translation can still cause mistakes, sometimes even creating some humorous effects. One of the most cited examples that has also been widely circulated on the web is a photo of a bilingual sign with the Chinese 请在一米线外等候, which literally means "Please wait behind the one-meter line"; underneath the Chinese sentence is the English "translation": Please wait outside rice-flour noodle. This proves to be the drawback of statistical translation method. In Chinese writing, there are no spaces between characters and words, thus making word segmentation by the computer extremely challenging. In the expression 一米线 (one-meter line), both 一米 and 米线 can be analyzed as possible words in Chinese. For this sign, it is possible that the translation software automatically selected 米线 without the pre-modifier 一 (one) because 米线 (rice noodle) is more frequently occurring than一米 (one meter) in the corpora. While it is still uncertain whether the image is authentic or if it is just an internet hoax, my attempt to Google-translate this sentence gives a similar result: Please wait outside in a noodle (see Figure 7.1). This rice noodle case is discussed by Mair (2011) in greater detail in Language Log, an American linguistics blog (http://languagelog.ldc.upenn.edu/nll/?p=3366). Similar machine translation errors or humor also become the subject of much online discussion, and there are websites devoted to them, such as the Google Translate Fails page on Facebook, which posts images of funny google translate fails (www.facebook.com/GoogleTranslateFails/?fref=ts).

As an example of media convergence in the age of Web 2.0 social media, online translators are no longer independent tools; they can be incorporated into social network sites. For example, Facebook now allows users to view foreign language posts in their native languages through the "see translation" function. Launched in 2011, this is an automated in-line translation service powered by Microsoft Bing. It appears as a link underneath a status update or a comment written in a language that is different from the default language of one's internet browser. When the reader clicks on the "see translation" link, the message is instantly translated into the default browser's language, and the user can also rate the level of comprehensibility of the translated text from a scale of very good to very bad. The actual effectiveness of this feature, which is regarded as part of Facebook's effort to facilitate global communication, has been questioned and criticized. Hendus (2015) sees this feature as one of Facebook's strategies in strengthening and expanding its global market. By imposing an explicit multilingual policy on its users and by gradually introducing features that claim to encourage multilingualism and diminish language barriers, Facebook is also influencing users' actual linguistic practices and language choices on the site. Eventually, the

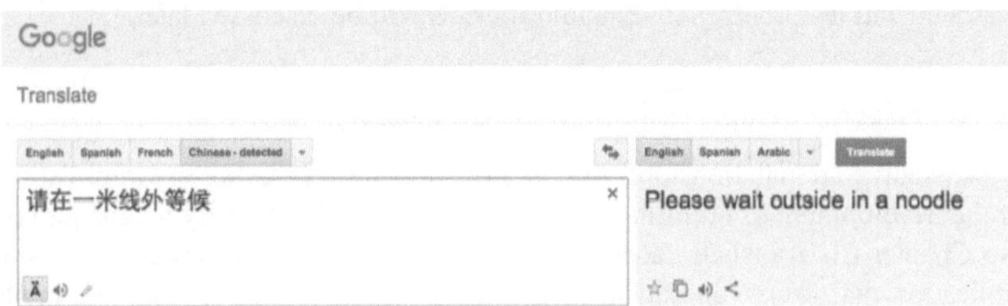

Figure 7.1 Please wait outside in a noodle, translated by Google Translate (© 2015 Google Inc., used with permission. Google and the Google logo are registered trademarks of Google Inc.)

more multilingual possibilities it offers, the more likely it will appear as a truly global site that is "available to everyone in the world" (Facebook, 2016).

COMMUNITY TRANSLATION AS A VERNACULAR MULTILINGUAL PRACTICE

As mentioned, both Google Translate and the "see translation" function on Facebook rely on machine translation rather than supervised human translation. This means that their translation results are based primarily upon the software, and users play a rather passive role in the translation process. The affordances for self-generated contents in Web 2.0 social media have presented new opportunities (as well as challenges) for translation practices and research. *Community translation* has emerged in social media as a new form of multilingual practices involving nonexpert translators. Basically, community translation is about "general Internet users acting as translators" (O'Hagan, 2011: 11) and is broadly defined as "a wide range of [t]ranslation, carried out based on free user participation in digital media spaces where Translation is undertaken by unspecified self-selected individuals" (O'Hagan, 2009: 97). Other labels that have been used interchangeably with community translation include *crowdsourced translation, collaborative translation*, and *social translation*. One of the objectives of community translation is to gradually "teach" the web different languages through self-generated translations provided by web users themselves. Community translation is also a collaborative effort, involving general web users working as affinity groups or "crowds" of translators on specific projects. That is why it is sometimes referred to as *social translation* (Pérez-Ortiz, 2011). Many community translation projects are initiated by global companies and organizations, such as the TED Open Translation Project, which invites web users to provide subtitles of TED talks in different languages (Olohan, 2014). Other examples include Duolingo and Facebook's translations applications. Some community translation projects are not entirely self-generated by users. As will be discussed, projects like the Facebook translation and Duolingo are actually controlled by a top-down language policy, as they are initiated by organizations with their intent to elicit volunteer translators from the crowd instead of hiring professionals (Lenihan, 2011, 2013; Hendus, 2015). Other forms of community translation, such as fansubbing, are more user-generated and less controlled by authorities, as will be discussed later.

Translate Facebook app and minority languages

Crowd-sourced translation applications and practices have important implications for the development of multilingual resources, especially minority languages (see also Chapter 6). Translate Facebook (also known as the Translations app) is an application that users can add to their Facebook profile pages. Once the application is added to one's Facebook profile, the user can then become a translator of any language of their choice. Translations of any content on Facebook can be submitted to the community of translators, who will then review and approve of the translations through a voting system. For Facebook, their ultimate goal is to "translate Facebook into every language in the world" (Facebook Site Governance, 2009, cited in Lenihan, 2011: 56). At first glance, this initiative seems to promote linguistic diversity. It may even help revitalize some languages that used to be unwritten or under threat. An example in point is Irish. Irish has its own community of translators on Facebook. Despite its official status in Ireland and the EU, it is still considered an endangered language with its decreasing number of regular users (Lenihan, 2011), as discussed in the previous chapter. While the application has opened up new multilingual possibilities, it is still governed by Facebook's top-down language policy and policing. Lenihan (2011) argues that the application does not automatically bring about true linguistic diversity. First, its translation page interface is available only in English; second, translations can only be submitted via the US English site; third, Facebook administrators can only receive and answer feedback or questions in English. In fact, this problem is not only present in the Translate Facebook application; all other requests sent to Facebook officials have to be communicated only in English.

There are also enough reasons to believe that such applications as Translate Facebook are global companies' "exploitation of the Internet crowd to obtain free translations . . . by going outside the professional translation sphere" (O'Hagan, 2011: 14). The Translate Facebook application is yet another strategy of Facebook to market itself as a global company that appears to embrace multilingualism, or what some refer to as *fake multilingualism* (Kelly-Holmes, 2005; Lenihan, 2011, 2013). A similar practice is adopted by Flickr. Every time a user logs on to Flickr, it greets them in a new language ranging from English to Chinese to smaller languages such as Mäori, an indigenous language in New Zealand spoken by just a few thousand people. Another reason why Translate Facebook practices a limited form of multilingualism is that each language and its varieties have their own translation applications pages that are separated from other languages. As Lenihan observes, this implies that Facebook still exercises a rather traditional view of multilingualism and diversity, where languages are seen as independent entities (see also discussion of the meaning of multilingualism in Chapter 1). Certainly, how social network sites discursively construct their global images through multilingual translations, whether artificial or natural, would be one of the key directions for future research on web-based multilingual practices.

Nonetheless, the Translate Facebook application does make minority languages visible at the bottom-up level. Returning to the translations page for Irish, a closer look at the metalinguistic discourses of the Irish community of translators revealed their commitment and heavy involvement in maintaining their language. For example, through discussing and evaluating one another's translations, members managed to translate and discuss words and phrases that remained untranslated elsewhere. As Lenihan (2013) noted in her analysis of metalanguage among the Irish Translations community on Facebook:

> The terminology developed by the community via the Translations app may aid the uptake of minority languages by other new media entities and users as it provides terminology relating to the online world. . . . This could consequently increase the "language footprint" of minority languages, making them more visible and their

> communities gain new space(s) and opportunity to use the language in other new media domains. . . .
>
> Lenihan (2013: 369–370)

In addition to Irish, social translation online may have facilitated the maintenance of other minority languages such as Catalan. Catalan is a native language of Catalonia, an autonomous bilingual region in the northeast of Spain. It is often recognized as a minority language due to its regional usage and decreasing population able to read and write it. In an early study, Climent et al. (2003) expressed concerns over the challenges in maintaining Catalan with the rise of the multilingual internet. They offered a detailed analysis of a corpus of Usenet posts in a Catalan-language university and found that 76% of the posts were written in Catalan. However, there was a significant tendency of shifting to Spanish in spontaneous replies to previous postings by non-Catalan speakers. Such strong preference for Spanish, according to Climent et al., could pose potential threats to the future of Catalan. The researchers, therefore, carried out a detailed analysis of the structural features of the Catalan posts, with the hope to develop a translation system between Catalan and other languages. Their ultimate goal, as with many other translation efforts, was to encourage Catalan speakers to be at ease in using their native language even in cross-linguistic communicative contexts. They concluded that "the future of Catalan depends on its users". With the advent of user-generated translation on Web 2.0, Catalan may have indeed been saved by its growing community of translators on the internet, among other factors. The Catalan edition of Wikipedia, for example, now contains over 400,000 articles, making it the fourth largest amongst Italic languages, a branch of the Indo-European language family (Wikipedia, 2015). At the same time, a large number of Catalan fansubbed videos, especially those of Japanese animes, have been shared and discussed on YouTube. What we are witnessing is that what used to be labelled as minority or endangered has now made its way to the global mediascape.

Fansubbing

Compared to the top-down community translation practices discussed in the previous section, *fan translation* is typically *voluntary* and unaffected by explicit language policies. The idea of fan translation has its origin in video gaming, where fans of Japanese video games translated the textual interfaces or instructions of games from Japanese into English so that they could be playable by gamers who do not understand Japanese. A more recent trend in fan translation is *fansubbing*, that is, fans providing amateur translation for the videos of their favorite TV series, animes, movies, music videos, and so forth. The translated videos are often published in the form of subtitled videos posted on video-sharing platforms such as YouTube and on blogs. Many fansubbers consider fansubbing more as a hobby than as real work, as what they do is "unpaid, decentralized and self-organized" (Lee, 2011: 9). A similar fan translation activity is called *scanlation*, in which fans scan pages of Japanese manga or comic books into digital editable images; the original Japanese texts can then be replaced by texts in a translated language by using photo-editing software such as Photoshop. Fansubbing, like fan fiction, is an excellent example of digital mobilities and superdiversity. Not only does it encourage flows of linguistic resources, but the global reach of fansubbed videos is unprecedented. Some popular videos are available in a wide range of languages, including minority and even previously unwritten ones. One surprising example that I recently came across is that the theme song of the popular Disney movie *Frozen*, "Let It Go", has been translated into at least 26 Chinese dialects (see: Let It Go [26 Chinese dialects + English translation] on YouTube: https://youtu.be/uaz1rEoJTtk).

To date, fansubbing has received little attention in linguistics research. An exception to this is Benson and Chan (2010), who examined fan subtitling of the 2008 official Olympics song "Beijing Welcomes You" on YouTube. The video has been translated and fansubbed into English and other languages such as Spanish and Vietnamese. As a form of "multilingual digital literacy practice" that involves a great deal of "language work" (Benson and Chan, 2010: 5), fansubbing offers new informal learning opportunities for fansubbers to improve on their linguistic knowledge; they do this through translating, subtitling, and editing the original videos. The associated practices of fansubbing also contribute to the circulation of metalinguistic discourses. As many fansubbed videos are posted on platforms with commenting spaces, they provide a public venue for native speakers of the subtitle languages to evaluate the accuracy and quality of the fansubs. Positioning themselves as experts of their native languages, as Benson and Chan note, the commenters often come up with alternative or what they consider to be better translations. For the "Beijing Welcomes You" video, the commenters who discuss the translations may be learners of Chinese who take up the open-ended commenting space as a space for informal learning and sometimes even teaching. In the comment thread shown in Example 7.1, a mini Chinese lesson takes place between a few Chinese learners and a "half-ish" Chinese commenter ("diaryanjo").

Example 7.1 **A mini Chinese lesson**

oxTwilightxo100: ni hao'..that means hello in chinese im takn chinese in school
fredastairefan: Me too!!! . . . Hi hao ma? Wo hen hao!
francis2383: hmmm hi HI must be Ni
diaryanjo: uh!! noo!! Ni hao translated actually means you good? Ni means you. . . . . Hao means good!!! but it is mostly used to regard a person saying hello. . . . . . . any more questions?? concerns?? i'm not chinese but i am still half!! ish..

(Benson and Chan, 2010: 16)

Example 7.1 is a case of incidental learning in that there is no clear discourse of deliberate teaching in the comment. Although the fansubbed video was not intended for any educational purposes, the viewers apparently took up the affordances provided by YouTube's commenting space to exchange their knowledge of Chinese. In terms of language choice, the insertion of pinyin Chinese in these comments is an assertion of learners' knowledge of Mandarin Chinese. Initiated by "oxTwilightxo100," this thread is centered around greetings, a rather basic yet typical topic in any elementary language lesson. Through discussing the literal translation of *ni hao*, meaning "hello" or "hi", the commenters self-generated a fruitful lesson of basic Chinese greetings from an insider learner's perspective ("im takn Chinese in school") or people with an ethnic Chinese background ("i'm not Chinese but I am still half!! ish").

The cases of fansubbing or fan translation more generally illustrate the ways in which the practice of translation has been transformed into a more complex nexus of *vernacular multilingual practices* (Barton and Lee, 2012, 2013). Vernacular literacy practices refer to reading and writing activities that are rooted in people's everyday experiences (Barton and Hamilton, 1998). In the context of this book, I am referring to vernacular literacies involving the use of multilingual resources online. These literacies are not skills but are embedded in what people *do*. My use of the term *vernacular* here is different from the traditional sociolinguistic notion of vernacular language, which usually refers to a native variety or dialect belonging to a specific group. Vernacular literacy practices are not necessarily tied to specific languages, especially on the global internet. An important feature of vernacular literacies is that they are voluntary and self-generated, rather than being framed and valued by the needs of social institutions. In the pre-web era, handwritten diary keeping used to be a common form of vernacular literacy in that it was a private and nonmainstream writing activity. There are many overlapping areas between forms of participation in community translation in digital media and the characteristics of vernacular literacy practices. Barton and Lee (2012, 2013) have outlined the features of vernacular literacies on Web 2.0:

- They give rise to new affordances and practices.
- They are self-generated.
- They are sources of creativity, invention, and originality.
- They are valued publicly.

Practices associated with crowdsourced translation are clearly vernacular literacies in the digital age. The translation activity often leads to the discovery of newer affordances and practices. We have seen communities of translators take up the affordances of YouTube to share and discuss their knowledge of languages, thus giving rise to new learning opportunities. Fansubbing practices are mainly voluntary and are not regulated or supervised by any authorities. The activity of fan translations often gives rise to creativity and originality. In the case of the Translate Facebook application, translators of minority languages have made visible language resources that had never been translated elsewhere. In discussing their translations, the translators are at the same time producing different sorts of language ideologies, which, in one way or the other, influence the development and maintenance of the languages in question. Contrary to older forms of vernacular literacy practices, such as diary writing or note-taking on paper, vernacular literacies in digital media are more valued and more widely circulated by the networked public.

GETTING THINGS DONE THROUGH TRANSLATION PRACTICES IN THE MOBILE WORLD

It becomes clear that using translation applications or actively translating contents online is not just about working out the meaning of words and phrases in different languages; more importantly, it is about getting things done in different domains of life (Barton & Hamilton 1998; Lankshear and Knobel, 2011). The invention of smartphone technologies and mobile apps makes it easier to embed translation activities in people's lived experiences, both online and offline. There are ample examples of how people around the globe getting things done with Google Translate. With its automatic speech detection technology, the Google Translate app makes possible conversations between two or more mutually unintelligible speakers. It is not uncommon for tourists in a foreign country to use a translation app on their smartphones when asking locals for directions or communicating with a shop assistant who does not speak the tourist's language. I have once used the app to get myself some painkillers in a pharmacy in Japan (and it worked). These new practices of translation are also changing the practices of intercultural communication. In view of these, the online and the offline dichotomy is also diminishing with the increasing use of apps and smartphones to carry out "physical" activity. Such practices are often multimodal, transmedia, and transcontextual (Androutsopoulos, 2015). In the following, I give a few examples showing how online translation facilitates the work of nurses in Japan, enhances multilingual participation on Flickr, assists the circulation of information during a political movement, and gives rise to acts of deliberate learning.

An interesting academic study which illustrates the intertwining relationship between machine translation and people's lives is Anazawa et al.'s (2012) study of how Japanese nurses can benefit from Google Translate at work. For their professional development, nurses and nursing students in Japan are gradually required to read international research literature published in English and other foreign languages. With limited proficiency in these languages, they are increasingly reliant on free translation tools such as Google Translate. However, Anazawa et al. also found that some nurses in their study experienced problems in understanding the translated texts, especially those from English to Japanese. This of course has implications for the development of machine translation technologies.

Translation is also an important multilingual practice on Flickr. In Lee and Barton (2012), Google Translate was mentioned by several participants as an important tool that facilitated new multilingual encounters in their Flickr participation. Even though some of them had limited knowledge of English, they made an effort to translate content from their local languages into what they consider to be a more global language, English, for their international audiences (Lee and Barton, 2012). The following excerpt comes from the profile of an Italian Flickr user called "Angelo":

> My English is Google translator. . .
> Mi español es el traductor de Google. . . .
> Mon français est traducteur de Google . . .

This translated text, which is suggestively humorous, serves multiple social functions: First, it allows Angelo to playfully position himself as someone who knows limited English, Spanish, and French and thus is seeking support from other Flickr members; on the other hand, this mixed language text illustrates what Jørgensen (2008) calls *polylingual languaging* – playful deployment of linguistic resources regardless of linguistic competence. Similar to the use of foreign languages in fan fiction writing, such playful deployment of languages enables Angelo to perform an international identity and widen his Flickr participation (see also Chapter 4). But unlike fan fiction, Angelo's multilingual resources in this instance were mediated by Google Translate. Advances in free online translators have certainly presented new multilingual experiences and changed the way people engage in translation activities in society. Furthermore, many Flickr users also actively participate in translation by drawing on their available multilingual resources. Examples include mixed language writing in usernames, profiles, tags, and captions. These are not simply instances of code-switching; a large number of tags, for example, are *parallel bilingual texts*, which involve literal translation from one language into another (Sebba, 2012). Doing so enables users to assert a more glocal and cosmopolitan identity, thus widening their participation in the globalized Flickr community.

Social network sites also serve as mediators of translation practices between the so-called online and offline worlds. This is evident in the various self-organized online networks for political movements. During the umbrella movement in Hong Kong (also discussed in Chapter 4), social network sites played a significant role in the organization of events and reporting of happenings about the movement. In particular, news originally reported in Cantonese or printed in Chinese in various local mass media was translated by voluntary web users into English and other languages. The translated texts were then edited and shared on global newsfeeds such as reddit, Twitter, and Facebook. Some of these were student-led initiatives. For example, a group of international students living in Hong Kong started a community page on Facebook called "Translating the Umbrella Movement". Their aim, according to the description on their page (Example 7.2), was "coordinating volunteer translators and foreign media outlets on the ground for greater international coverage of the current situation in Hong Kong" (About Translating the Umbrella Movement, Facebook, 2014–2015). What is interesting about this community page is that about 90% of the posts are available in English only; posts that are in Chinese are articles and information linked from external resources. In fact, English seems to be the preferred base language for foreign language translations. As suggested in one of their calls for translators, (see Example 7.2) they were looking for people who were able to translate news "from Cantonese to English to any other language". Therefore, except for Cantonese to English translators, other translators were mainly involved in what is referred to in translation studies as *indirect* or *second-hand translation activity* (Aranda, 2007).

Example 7.2 **About "Translating the umbrella movement"**

Please let us know if you are willing to translate from Cantonese to English to any other language so we can put you in contact with the media outlets on the ground!

12.22pm Desperately in need of a Spanish speaker!
12.52pm Found some Spanish speakers! Thank you!
16.10pm Really need a lot more CANTONESE to ENGLISH translators!*
16.57pm More Cantonese/English speakers needed!
We now have the following translators available:
- English
- Cantonese
- Mandarin
- Japanese
- French
- Spanish
- Basque
- Bengali
- Russian
- Turkish
- German

This suggests that the creative aspect of this Facebook community is not so much about the actual translation per se, given that most of the posts are not translated texts; rather, through this page, the organizers took up the affordances of social media to drive and influence translation practices in the offline, physical protest sites on the ground. The page successfully brought together protesters, multilingual translators, and foreign journalists who were out there on the protest sites but might have problems understanding Cantonese and written Chinese in print media. This thus created an ad hoc network of translators (Bruns and Burgess, 2014), in that the page and the translators were only needed at the time of the umbrella movement. After the movement, the page was only updated intermittently and, as of July 2016, the page had not been updated since June 2015.

Finally, fan translation such as fansubbing has been taken up as a space for informal and collaborative language learning. In addition to Benson and Chan's (2010) study on the metalinguistic comments on YouTube fansubbing discussed earlier, Ito et al. (2010) also describe fansubbing as a space for learning and for young fansubbers to engage in cross-linguistic collaborations. A more recent study is Lakarnchua (2015), which surveyed 86 Thai fansubbers about their participation in multilingual fansubbing. First, the study shows that most participants were involved in second-hand translations, that is, instead of translating a Korean video directly into Thai language, they fansubbed the English translation of the original Korean version. It is, however, through such indirect translation that the participants were able to improve their English vocabulary. Lakarnchua also believes that the multilingual activity involved in fansubbing presents new opportunities for language learning. The following comments reveal the fansubbers' implicit theories of language learning:

> Making subs is a form of language practice in itself. There are both vocabulary words that I know and do not know. Any vocabulary words that I do not know, I go find the meaning of. Seeing the word in passing and seeing it frequently, I'll remember it. (Respondent 5)
>
> It helps me to practice my translation skills and I'm always learning new vocabulary words. (Respondent 9)
>
> Because it's doing something which we have an interest in, it will help us to absorb vocabulary words more quickly and help us to understand them better than just memorizing them. (Respondent 16)
>
> (Lakarnchua, 2015: 10–11)

These responses suggest that fansubbing not only gives rise to incidental learning, as Benson and Chan's (2010) study suggests, but it also provides a space for *deliberate acts of learning* (Barton, 2012). Deliberate learning, often contrasted with unconscious or incidental learning, takes place when learners actively and deliberately take agency and control of what they consider to be a learning activity. All of the responses cited point to the fact that fansubbing helped learners discover something new about language. Deliberate learning is particularly evident in respondent 5's comment: "Any vocabulary words that I do not know, I go find the meaning of". As a result, any deliberate act of learning is "a consciously reflective one" (Barton, 2012: 143), through which learners actively reflect upon their learning process and progress.

Ordinary web users are also increasingly involved in *evaluating* the quality of translations. In October 2015, Google posted a video advertisement for Google Translate. The advert tells a story about how a boy's dream has come true with the help of Google Translate. Not speaking any English, the boy, whose name is Alberto, and his family moved to a small town in Northern Ireland. Hoping to join the local football club, Alberto managed to communicate with his coaches with the help of the Google Translate smartphone app. A video of Alberto's "life-changing story" was also posted on YouTube. This immediately stimulated a debate about the authenticity of this story and whether this was yet another ad by Google. Many of the commenters focused on the "heavy Irish accent" of the coach and the "really weird" Spanish used by Alberto. They also doubted if the app could recognize either of the accents (see Example 7.3).

Example 7.3 Comments on the YouTube video "Google Translate: Alberto's Story"

Haha no chance Google translate could understand someone with a northern Irish accent

> I REALLY REALLY REALLY doubt Google translate can understand that Spanish. And that English, for that matter.

What is interesting here is not whether Google can understand these accents, but the way in which ordinary people collaboratively evaluate translation technologies based on their everyday experiences as both language users and technology users. Whether it is Google Translate, Translate Facebook, or fansubbing, these tools all facilitate the circulation and distribution of language resources and ideologies about them around the globe.

THE FUTURE OF ONLINE TRANSLATION PRACTICES

In this chapter, I have discussed online translation in terms of multilingual practices. That is, as people use translation apps to get things done in their lives, they are at the same time engaging in multilingual activities. As they participate as communities of translators in social media and apps, they are also involved in networked multilingual practices (see Androutsopoulos, 2013a, on the concept of "networked multilingualism"). These networks of translators provide important examples of participatory culture in the digital age. The fact that the formation of such groups is often interest-driven fits in with what Gee (2004) describes as "affinity groups". Translation, at least from the examples I have discussed, is not exclusive to professional translators or language experts. As ordinary web users have access to the required technological affordances, they can exercise their power as native speaker users of languages. They also contribute their linguistic knowledge to the web collaboratively, with the shared aim of improving mutual intelligibility on the web.

However, as indicated in the case of Facebook translations, software companies appear to still bc taking a onc-languagc-at-a-timc approach to multilingualism. In translation, languages are often treated as distinct entities, and technologies are not designed for multilingual CMC texts that are often characterized as hybrid and heterogeneous linguistically and stylistically. Common features of multilingual CMC, such as code-mixed writing and playful languaging (such as the Taobao style writing cited in Chapter 4), may not be recognizable by Google Translate. In the case of spontaneous forum posts written in Catalan (Climent et al., 2003, 2007), and perhaps for all informal CMC messages, features such as intentional code-mixing and intentional deviation from spelling and grammar may pose real problems for the machine, which is not prepared for such "noisy input" (Raley, 2003; Climent et al., 2007). Raley (2003) thus proposed that the future of machine translation relies on a universally readable "networked Global English", that is, a network of Englishes including Netspeak or what he refers to as "netlish" – English on the internet. However, this is a simplistic view of Netspeak, that CMC is "structured according to certain formal features that emphasize abbreviation, a minimalism of expression, and the basic taxonomic unit of information, such as the list and the short paragraph" (Raley, 2003: 303). In CMC, however, there exist more variations than standards. As numerous studies in the CMC literature have already indicated, CMC is hardly one single variety of language, and therefore it is unrealistic to assume that machine translation can decode the highly individualistic and often creative CMC messages found across the internet.

There is also the question of demand and supply. As Cunliffe and Harries (2005: 166) note, "the provision of a translation facility can only really work if users have faith in the translations provided". Despite ongoing efforts to promote multilingualism and the use of native languages, not all web participants are willing to use their native languages in online interaction. As evident in interviews about language choice in the Flickr research reported in Chapter 2, discourses of English as the language of global communication are still prevalent on the web. Many users still consider English as their preferred language of intercultural communication online (see Chapter 2, Barton and Lee, 2013). Besides, whether web users take up translation tools depends largely upon their "media ideologies" (Gershon, 2010; Hendus, 2015), which are a set of beliefs users develop to "explain perceived media structure and meaning" (Gershon, 2010: 3). This concept also explains why there exists a discrepancy between Facebook's intention to bridge language barriers and users' lack of interest in translating. A Facebook user survey conducted by Hendus (2015) revealed that people were indifferent or even opposed to the "see translation" function. Not all Facebook users find the "see translation" function appealing because of their differing beliefs and perceptions about Facebook as a medium. Some said they simply ignored posts in languages they could not understand, as they thought that these posts were not intended for them to read. While Facebook seems to be promoting linguistic diversity by introducing multilingual applications such as Translate Facebook, it does not match with users' actual Facebook practices and their expectations of how Facebook should be used.

Community translations done by real people may help address some of the problems with machine translation, such as the translation of idiomatic expressions. However, there have been concerns over the quality control of crowdsourced translations (Anastasiou and Gupta, 2011; Giles, 2012). An oft-cited issue is that these crowdsourced translators are inexperienced amateurs who take up translation as a hobby. Accredited translators have expressed concerns over the rise of crowdsourced translators (O'Hagan, 2011). Brabham (2012), however, argues that "the crowd of amateurs in crowdsourcing, it turns out, is a pervasive myth" that has been discursively constructed by the press. By contrast, a fact that is seldom acknowledged in mass media is that "crowds are largely self-selected professionals and experts who opt-in to crowdsourcing arrangements" (Brabham, 2012: 394–5). The ongoing tension between professional translators and the crowds is also evident in online protest groups. "Leave Translation to Translators!" is a public group on Facebook which claims to be a "group for all those who believe that translations should be done by translators rather than computers or people who think they can do it" (Leave Translation to Translators!, Facebook, 2015). Gee and Hayes (2011) characterize such ongoing debates in the digital age as the "crisis of expertise". They argue that this is a result of people's lack of trust in experts and that professionals "undervalue what they do not know and overvalue what they do" (Gee and Hayes, 2011: 44). O'Hagan (2011: 21) suggests that:

> While professional translators generally do a wonderful job, it is time to recognise that their operating environments are significantly shifting, giving rise to new ways of working. By being better informed of the nature of the change and by becoming more reflective of current practices, the translation profession will be better served and more likely to survive and flourish in the long run.

This changing meaning of translation in the mobile world also implies that translation research is becoming increasingly interdisciplinary. Studies of machine translation and community translation must go beyond investigating the technicalities of software development. Researchers should take into account the actual practices and media ideologies of all stakeholders, including the translation professionals, online communities of translators, language policy makers, as well as users of translation services online.

8

RESEARCHING MULTILINGUALISM ONLINE

Current trends and future perspectives

Overview

- An overview of methods
- The researcher's role in online multilingualism research
- The multilingual researcher
- Where we are now and the way forward

AN OVERVIEW OF METHODS

CMC researchers have employed a wide range of methods for data collection and analysis. In this final chapter, I focus on some of the methodological approaches that have been adopted in existing research on multilingualism online. While most of the topics and issues have been covered in discussions of general CMC methodologies, this chapter highlights some of the issues that are specific to researching multilingualism on the internet. To date, no frameworks have been developed specifically for the study of online multilingualism. In principle, research designs that are applicable to general CMC research may also be suitable for researching online multilingualism. (For general discussions on CMC methods, see Herring, 2009; Barton and Lee, 2013; Androutsopoulos, 2013b; Bolander and Locher, 2014; Page et al., 2014). Existing studies of online multilingualism have covered a rich array of languages, CMC platforms, and research contexts. Qualitative, quantitative, and mixed methods have been used in this body of work. Some scholars have grounded their studies in the theories of traditional sociolinguistics, discourse analysis, and pragmatics, and they have adopted the analytical methods used in these fields accordingly. Some of the more common methods of analysis include conversation analysis, critical discourse analysis, and interactional sociolinguistics. Others have employed methods that are specifically designed for CMC research, such as computer-mediated discourse analysis (CMDA), web content analysis, technobiographies, and discourse-centered online ethnography (DCOE).

Table 8.1 is a summary of the methodological approaches adopted in selected studies of multilingualism in CMC. While it is impossible to include all of the studies in the area, the summary covers a broad range of CMC platforms, data sources, languages, and themes (see also Chapter 3 for a summary of research on code-switching). Some of these studies have already been discussed in previous chapters.

In terms of data sources, as indicated in Table 8.1, both screen-based and user-based data have been collected in previous research on multilingualism online (Androutsopoulos, 2013b). *Screen-based data* refer to textual outputs on the screen produced by online users in their online participations (such as blog posts and Facebook comments), while *user-based data* are those generated through the researcher's contact with the participants, such as interviews. Many researchers consider both screen-based and user-based data in their studies (Lee, 2007a, 2007c, 2012; Deumert and Masinyana, 2008; Leppänen et al., 2009, 2011; Das and Herring, 2016). For example, Deumert and Masinyana collected SMS texts and interviewed users, while Das and Herring combined data from conversation logs, interviews, and participant observation.

Even within the broad category of multilingualism, there are various subthemes and directions, each of which has its own methodological design. For studies focusing on measuring linguistic diversity on the web, such as those discussed in Chapter 2, quantitative surveys and distributional analysis of languages have been the major method (e.g. Kelly-Holmes, 2004; Hendus, 2015). Kelly-Holmes (2004), for instance, adopted survey questionnaire as a major instrument in understanding language choice among more than 2,000 college students from eight countries. For research on language choice, the aim is to identify and count the languages that are present on the research sites, mostly through content analysis, and statistics may also be generated (e.g. Callahan and Herring, 2012). Other studies are interested in the unique Netspeak features in CMC in a certain language, and these often involve counting of features or descriptions of illustrative examples, including my early research on the linguistic features of email and ICQ in Hong Kong (Lee, 2007a, c). For those interested in people's practices, ideologies, metadiscourse, and issues related to identities, qualitative and ethnographic methods are preferred (e.g. Lam, 2000; Androutsopoulos, 2006). As can

Table 8.1 Research methods in selected studies of multilingualism online

Author(s) (year)	*Platform (s)*	*Text Data*	*Method(s)*	*Language(s)*	*Theme(s)*
Androutsopoulos (2006)	German-based diasporic websites	Forum posts, interview data	Online ethnography (persistent observation, interviews)	German/Greek, English	Code-switching, identities
Callahan and Herring (2012)	University websites from selected countries	Website texts	Content analysis (longitudinal)	Multiple	Language choice
Chun and Walters (2011)	YouTube	YouTube videos of Wonho Chung's Arabic-English bilingual performances	(Critical) discourse analysis	Arabic, English	Orientalism, Parody
Das and Herring (2016)	Orkut	Scrapbook conversations on Orkut (greetings)	Computer-mediated discourse analysis, interviews, participant observation	Bangla	Politeness, social distance
Deumert and Masinyana (2008)	SMS	SMS messages, interview data	Frequency count, interviews	isiXhosa, English	Homogenization of SMS language
Ensslin (2013)	Wikipedia	Text of Wikipedia's multilingual policies	Critical and multimodal discourse analysis	Multiple	Language policy
Hendus (2015)	Facebook (see translation)	Survey	Survey	Multiple	Language policy
Herring et al. (2007)	LiveJournal	Texts on LiveJournal	Language coding, network analysis	Russian, Portuguese, Finnish, Japanese	Language networks
Kelly-Holmes (2004)	Internet in general	Survey and log data by respondents from eight countries	Questionnaire survey	Multiple	Language choice, linguistic diversity

Kytölä (2013)	Forum	Posts on Finland-based football forums	Discourse-centered online ethnography	Finnish, English	Metapragmatics
Lam (2000)	Email, chat	Log data, field notes	Discourse analysis, ethnography (participant observation, interviews)	Chinese, English	Identities
Lee (2007a, 2007c)	Email, ICQ	Email and ICQ logs, interview data	Frequency count, content analysis	Cantonese, English	Linguistic features
Lee (2012)	Facebook	Technobiographies, Facebook posts	Technobiographic interviews, content analysis	Cantonese, English	Identities
Lenihan (2011)	Facebook (Irish translations application)	Facebook posts and comments	Virtual ethnography, "lurking"	Irish	Language ideologies, metalanguage
Leppänen et al. (2009, 2011)	Blogs, forums, fan fiction	Log data, interviews, photographs	Mixed (mainly ethnographic)	Finnish, English	Translocality, identities
Lexander (2012)	SMS	SMS messages, interviews	Literacy practices analysis	Wolof/Pular, French	Literacy practices
Siebenhaar (2006)	IRC	Corpus of IRC log data	Dialect-standard ratio analysis	Swiss German, standard German	Code choice, language variation
Warschauer et al. (2007)	Email, online chat	Sample messages of email and online chat	Survey, interviews	(Romanized) Egyptian Arabic, English	Code choice, code-switching

be seen from Table 8.1, there is a growing tendency for mixed methods, such as combining surveys with interviews, discourse analysis with ethnography, and so forth (e.g. Deumert and Masinyana, 2008; Kytölä, 2013). Many of the studies also involve the researchers going back and forth between texts and practices (Barton and Lee, 2013).

These research directions and approaches are in line with general observations about CMC research – that methods adopted in online multilingualism research correspond to the three main strands or "waves" of research identified by Androutsopoulos (2008, 2015). The first strand of CMC research aims to identify Netspeak features that are specific to CMC media, and various chapters in *The Multilingual Internet* book worked toward that direction (e.g. Lee, 2007b; Nishimura, 2007). In multilingual CMC research, the first wave was largely driven by grand surveys of linguistic diversity online (such as those outlined in Chapter 2). The second strand of CMC research moves beyond feature identification to variation. This body of work is largely informed by non-structuralist approaches to linguistics such as pragmatics, sociolinguistics, and discourse analysis. Some works on multilingualism online have also followed this research direction, such as Callahan and Herring's (2012) content analysis of multiple languages on university websites and Ensslin's discussion of Wikipedia's multilingual policies. A third wave of CMC research acknowledges the fact that people are "always on" and that digital media are embedded in participants' lived experiences. This is reflected in the growing amount of ethnographic and practice-based research on multilingualism in social media such as Leppänen et al.'s (2009, 2011) study of Finnish young people's multilingual practices in fan fiction.

While acknowledging the importance of describing details of textual features and patterns in CMC, a growing number of scholars have turned to ethnographic methods. Ethnographic data are useful in revealing the actual social activities and *practices* surrounding multilingual writing online, so as to understand better how and what people write online, as well as what they do with their multilingual texts. An earlier study that adopted an ethnographic approach to multilingual CMC was Lexander (2012). In the study, Lexander combined text analysis with ethnographic interviews to analyze multilingual SMS as literacy practices. In what follows, I describe in greater detail two methodological approaches that are of particular relevance to understanding multilingual resources and practices in the global world, online and offline. The first approach covers methods that pay attention to both texts and user-based data, or what Androutsopoulos (2008) calls discourse-centered online ethnography (DCOE); the second approach is *technobiography*, a method for understanding how online linguistic practices are situated in participants' offline lives.

Focusing on screen and people: discourse-centered online ethnography

Androutsopoulos (2008) calls for a "third wave" of CMC research that adopts what he calls *discourse centered online ethnography* (DCOE), an approach that "combines the systematic observation of selected sites of online discourse with direct contact with its social actors" (2). He goes on to explain that "DCOE uses ethnographic insights as a backdrop to the selection, analysis, and interpretation of log data, in order to illuminate relations between digital texts and their production and reception practices". This framework has developed from his work on German-based web environments. For example, one of his projects examines sociolinguistic styles and identity constructions on sites devoted to hip-hop culture (Androutsopoulos, 2007). In doing DCOE, the researcher often starts with systematic observation of the discourse of the sites, then interviews internet participants or elicits insiders' perspectives through other means. A study of online multilingualism that is informed by DCOE is Kytölä (2013). In the study, Kytölä began with his persistent observation of two discussion spaces on the Finland-based football forum Futisforum. Through signing up as a member of the forum (and being a football enthusiast), Kytölä was able to have informal contact with the forum participants, so as to understand better their metapragmatic discourses and multilingual participation. This was then complemented by the collection of selected forum posts and interactional sociolinguistic analysis.

The approaches I have taken in my own research on multilingual online practices are parallel to the premises of DCOE. The overall objective of my work has been to connect texts and practices. For instance, consistent with Kytölä's research, my collaborative research with David Barton on multilingual practices on Flickr (Lee and Barton, 2011) started with observations and the researchers' own interest in photography and Flickr. It involved identifying 100 photo streams and "persistent observation" (Herring, 2004) of them to obtain a snapshot of the distribution of languages in different spaces on Flickr (e.g. captions, tags, and comments). The next phase of the research involved an online questionnaire survey about users' general Flickr use, through which we also identified 30 focal participants (18 Chinese speakers and 12 Spanish speakers). This was then followed up by a series of online interviews via FlickrMail or private email with the participants, so as to identify different ways of participating in Flickr and ways in which these users deployed their linguistic resources on their own sites. The interview questions focused on specific areas of these Flickr sites (e.g. why English was used when describing a particular photo while Chinese was used in the tags) as well as the answers to the initial survey questionnaire. The interviews with active Flickr participants became the core data for the research as they often revealed details about actual situations of Flickr use, as well as differences among the participants' practices. The methodological design in the Flickr project is more or less in line with the mixed methods approach taken in DCOE, as Androutsopoulos (2008) explains:

> While interviews may offer insights that are not (or only marginally) accessible through systematic observation, observation may disclose aspects of structure that are difficult to elicit in an interview. At the same time, linguistic analysis of log data may contextualize emic views, indicating where participants' distinctions are generalised or biased.

In combining discourse analysis with a practice-based approach, it is important to note that neither text nor practice should be seen as the sole point of departure in research. Rather, researchers often need to go back and forth between language data and practice data to understand better the interplay between what is visible on the screen and what people actually do with their language resources online.

Focusing on lived experiences online and offline: technobiography

At the very beginning of this book, I told a story of my experiences with technology since I first learned to use the computer. I also mentioned how these experiences have shaped my multilingual practices on various online platforms over the years. This story constitutes what is called my *technobiography* (Henwood et al., 2001; Kennedy, 2003; Barton and Lee, 2013), which consists of "accounts of everyday relationships with technology" (Henwood et al., 2001). In this subsection, I first offer a general discussion of the term technobiography and its background. I then describe how this has been applied in online multilingualism research.

The term technobiography was first coined and applied extensively in a collection of women's life stories about their use of technology in Henwood et al. (2001) and was then further developed by Kennedy (2003). Like any biography, a technobiography always starts with a person, with the aim of finding out about who this person is through studying and telling a story about his or her life. These experiences are *historically situated.* In other words, a technobiography is located at different points in time. In the context of online research, the term technobiography may refer to a method or a product. As a method, it is one which allows the researcher to situate participants' online practices in their offline lives, and vice versa. Ching and Vigdor (2005) further developed the concept into a form of interview, or *technobiographic interview*, which focuses on the interviewee's encounters with technology "at various times and in various locations throughout their histories" (4). As a product, technobiographies are stories produced from people's accounts and narratives of the roles of technology in their lives. In addition, technobiography is not just about identifying which technology is used in what domain of life, but it is also about how people feel about the role of technologies in their lived experiences. In other words, technobiographies are highly reflexive in nature.

In my own research on multilingual practices and identity performances among university students in Hong Kong (Lee, 2014), I largely adopted a technobiographic approach to elicit data about the relationship between multilingual practices and identity performances. The questions that I devised in the technobiographic interviews closely followed eight key areas: current practices, participation, a day in life, life history, transitions, domains, comparisons, and focus on language. These categories were first used in courses taught by David Barton at Lancaster and were further developed in Barton and Lee (2013) and Page et al. (2014, 131–132).

Topics covered in technobiographic interviews

1. *Current practices:* What are the sites you use most often, and what are the ones you have contributed to?

2. *Participation:* In what way do you contribute contents online? Privately or publicly? Have you commented on news or products? Voted on the quality of service? Submitted a review or a Wiki entry? Uploaded pictures or videos for comment?
3. *A day in life:* Think of yesterday. What technologies did you first deal with when you woke up, and how did this continue during the day?
4. *Life history:* What was your first experience with computer technology? When did you first go online and what did you do? When did you first use a mouse, send a text message, search Wikipedia, start using Facebook, and so forth?
5. *Transitions:* Did you change your practices of keeping people's addresses, arranging to meet friends, using maps, etc.?
6. *Domains:* Do you use different technologies/websites in your everyday life, your student life, and in any work life? What about other domains, such as religion, sports, politics?
7. *Comparisons:* Are there differences across generations, parents, grandparents, children; differences across cultures, friends from other countries; gender differences, prohibitions? If yes, what are they?
8. *Focus on language*: Do you adopt an "e-grammar" style in online communication? Is your online writing different from your writing elsewhere? Have you changed your linguistic style online over the years? In what ways? If you speak different languages in your everyday life, how do you deploy these languages in your online writing?

These questions serve as prompts to facilitate participants' narratives about their technology-related lives, and they are by no means an exhaustive list. The last category, *focus on language*, is a newly added one which is especially relevant in multilingual CMC research. The questions listed under this item aim to elicit data about how offline language choice may shape online linguistic practices and vice versa. They also aim to trace changes in practices over time. In fact, eliciting narratives of participants has been a major method of data collection in various areas in linguistics, notably language acquisition and sociolinguistics (see, for example, Busch et al., 2006; Pavlenko, 2007).

A technobiography with a focus on language can also be referred to as a *technolinguistic biography*, a subtype of technobiography. There is no particular order in which these eight areas and questions appear in the course of a technobiographic interview session. Language-related questions can be interspersed among the other areas and questions. One possible drawback of a technobiographic interview is that it can be very large in scope because it covers a whole life history with technology. From my experience, some interviewees would spend more time on talking about technologies or their lives and reveal little with regard to their language use. That is why technobiographic interviews should be complemented by other methods and data sources. For example, my study was organized into two phases. Before the actual technobiographic interview, the participants were invited to fill out an online questionnaire that provided me with a snapshot understanding of the student participants' general online habits and their available multilingual resources on- and offline. In this phase, I also performed persistent observation of their most frequented sites reported in the survey. Then, from the people who agreed to further participate in the project, I invited 20 students to participate in the main phase of my study, which comprised a screen-recording session, followed by an in-depth technobiographic interview. In the screen-recording session, each participant was asked to go online for about 30 to 50 minutes on a computer in the researcher's office. Having reviewed the recording, the researcher was then able to come up with interview questions that were specifically designed for that participant.

To date, however, research that adopts a technobiographic approach has assumed that technobiographies are collected through interview data only. However, in reality, technobiographies can take many forms that go beyond interviews. Nowadays, most social media are profile-driven. In people's online profiles, their life stories are constructed by sharing pieces of their lives such as their hometowns, education, work places, and so forth. Another popular form of self-disclosure is achieved through continuous status updating. Status updates and posts about oneself on Facebook and Twitter are in fact small stories (Page, 2011). The multimodal affordances of new media also allow for more visual representations of technobiographies. For example, personal photos are organized into public online albums in social media, which can serve as important sources of technobiography. The technobiographic researcher needs to pay close attention to all these possible forms of technobiography to obtain a holistic view of a participant's relationships with technology. Having gathered all these data, the next step is to collate the information and write up the technobiography. The Appendix is the technolinguistic biography of one of my research participants, Tony. This example has also been discussed in terms of language choice and identity in Barton and Lee (2013) as well as Lee (2014).

A technobiographic approach to research is more than just an ordinary face-to-face or online interview. Technobiographic data cannot be elicited through just one or two interview sessions. Because it is about life, a comprehensive technobiographic interview can be longitudinal and ongoing. For example, Tony (see Appendix) was first interviewed in 2011; I have revisited his case in 2015 and conducted a detailed interview with him about changes in both his online and offline lives. For example, Tony's "working holiday" experience in Ireland in 2014 largely shaped his postings in English on Facebook (see the epilogue section in the Appendix). In other words, building a rapport with the participants is crucial to technobiographies. It is not just about what the participants say in the interviews, but also about the footprints they have left online and offline. That means technobiographic interviews are often based upon ongoing observations of the participants' online and offline activities. The researcher needs to regularly stay up-to-date on the participants' online profiles, posts, and so forth. The number of follow-up interviews may be unpredictable and can be carried out online or face-to-face. Because one of the aims of this method is to trace changes in life history, it is especially ideal for longitudinal research projects or those that trace the trajectory of people's digital practices.

So far I have focused on what technobiographies are in general terms. There are more specific reasons why a technobiographic approach is beneficial to multilingual online research or CMC research at large. First, it situates people's online experiences in their offline lives. As Kennedy (2003: 132) argues:

> Since experiences of the Internet, hypertext, and digital multimedia are necessarily *offline* as well as *online* phenomena, research into offline experiences of these digital technologies should form the context for developing a greater understanding of online experiences.

As explained in Chapter 1, online and offline are used here loosely as convenient labels. Technologies are socially constructed by people who do not simply live in an online world. We cannot fully understand a participant's online behaviors and practices without understanding who they are and what they do in the offline world. It is often people's offline encounters, experiences, feelings, emotions, practices, ideologies, attitudes, and social networks that shape their online linguistic practices, and vice versa. As revealed in my technobiography at the beginning of the book, it is often the people from the offline world who had an impact on my online use of language, such as my brothers who got me onto the computer world; as a graduate student in the UK, I felt the need to keep in touch with friends and family in Hong Kong and also new friends from Lancaster, and that is why I started blogging in English. Through technobiographies, researchers can discover differences between online and offline linguistic choices.

Second, technobiographies reveal not only what kinds of technology play a role in one's life, but also one's feelings and attitudes. When talking about their own lives in relation to technology use, the participants often revealed their opinions and stance toward a certain linguistic code. Many of these views may even reveal their emotions and attitudes. These metalinguistic comments, or what Kennedy (2003) calls "symbolic discourses and feelings" about technologies, are often useful in understanding what different linguistic resources mean in their technology-related lives (see also Chapter 5 for more about metalinguistic discourses).

Third, every technobiography is unique. The aim of conducting technobiographies is not to make generalizations about people's online lives. Rather, they reveal details about individual differences. This is especially relevant in researching internet multilingualism. Details about individual users' online and offline linguistic repertoires can definitely inform our understanding of the level of linguistic diversity on the internet.

THE RESEARCHER'S ROLE IN ONLINE MULTILINGUALISM RESEARCH

In this section, I discuss the significance of reflecting on the researcher's roles in carrying out multilingual CMC research. Turning to more qualitative and ethnographic approaches to multilingual CMC research also implies a much closer connection between the researcher and the researched. As with qualitative research on any subject, such practice-based or user-based approach to multilingual CMC *requires* the researcher to establish a close relationship with their data and research participants throughout the research process. It is a requirement because studying digital texts inevitably involves the researcher looking closely into and discussing personal, and at times sensitive, online communication with the participants. A sense of trust between the researcher and the research participants has to be established through ongoing rapport building in the research process. In addition, the unique features of online meaning-making and practices have called for more *insider research*, that is, research that relies on knowledge generated by those who are already actively involved in the research site.

Insider research can take two major forms. One is to elicit insider perspective through having direct contact with the participants, as with carrying out a DCOE. Common instruments for conducting this form of insider research include interviews and participant observation. The other form is research carried out *by* someone who is already an active participant of the research site. Some methods applied in CMC research include technobiographies and autoethnographies in which the researcher takes an active part in studying themselves (e.g. Davies and Merchant, 2009).

The importance of the researcher being an *insider user* in digital media research is especially favored by digital literacies scholars as in various studies of autoethnography (e.g. Davies and Merchant, 2009; Barton, 2015b). One reason for carrying out insider research is that the internet is naturally diverse, especially in thc meaning-making and interpretation processes. Lankshear and Knobel (2009: 9) point out that:

> [a]lmost anything available online becomes a resource for diverse kinds of meaning making. In many cases the meanings that are made will not be intelligible to people at large or, in some cases, to many people at all. Some might be shared only by "insiders" of quite small interest groups or cliques.

Another reason why an insider perspective is invaluable is that digital media are an indispensable part of many people's lives. Davies (2007: 250), when carrying out an autoethnography of her own literacy practices on Flickr, noted that "digital texts are ubiquitous, increasingly embedded in the lives of those who have adopted them" and, therefore, "insider knowledge is required" and can be a "real asset". As with doing ethnographic research on any site, multilingual CMC researchers need to familiarize themselves with what goes on in their research sites before they can study them. For multilingualism research online, this means more than just signing up for an account and "lurking". Researching multilingualism online from an insider perspective involves familiarizing oneself with the available linguistic resources. In my early IM research (Lee, 2007a), for instance, my familiarity with IM-specific language features in Hong Kong (such as Romanized Cantonese and a range of Asian emoticons) allowed me to compare my own experience and knowledge with my participants' practices. I was then able to discover the diversity in text-making practices among my participants.

As an insider analyst, understanding my participants' online multilingual practices through reflecting upon my own practices in and out of the research allowed me to understand both their texts and text-making practices in greater depth. If

I had never been an active user of Flickr who engaged in multilingual interaction, I would have never developed an interest in researching it in the first place; if I had never observed my contacts' Flickr sites and interviewed them, I would have never understood the diverse range of multilingual practices there. It becomes clear that researching multilingualism online goes beyond identifying or counting languages. What makes the research more meaningful is the specific in-group, insider knowledge of the research context in question, which can only be understood by having direct contact with the actual users.

The heavy involvement of digital researchers in their research contexts also means that the research project in question is co-constructed to some extent by both the researcher and the people being researched. Methodological issues such as ethics, representativeness, and objectivity naturally arise (See, for example, Herring, 2002; Page et al., 2014, for more extensive discussion of ethics in CMC research). It is thus essential for the researchers to be *reflexive* about their involvement and possible influence on the research participants and the research site. In ethnographic research, *reflexivity* refers to "a turning back on oneself, a process of self-reference" (Davies, 1999: 4). Papen (2005) further suggests that reflexivity in ethnographic research is more than just looking inward towards the research site, but also distancing oneself from it, taking a critical stance and even problematizing the methods adopted in the project.

Reflexivity in CMC research is not a new phenomenon. For example, Herring (1997), in an early discussion of ethics in CMC research, put forward the various problems that she had experienced in her research on gender in CMC. Though not discussing reflexivity explicitly, this is certainly one of the earliest discussions of CMC methods through reflecting upon the researcher's own experiences and problems encountered. However, to date, few CMC researchers have made their roles explicit in their published research, and not many have discussed the importance of reflexivity in CMC research (with few exceptions, for example, Chun and Walters, 2011, who reflected on their critical stance towards race on YouTube, and Lee and Barton, 2011, who carried out autoethnographies of their own participation on Flickr).

As with all kinds of qualitative research involving human subjects, it is essential for online multilingualism researchers themselves to make their relationship to the research site and participants, to their research project, and their stance very *explicit* in their analyses and writing. As well as getting close to my research participants, in my previous research projects on multilingualism online, I constantly reflected on my own participation, thus carrying out my autoethnographies, or to be precise, my autonetnographies, "autobiographical personal reflection on online community membership" (Kozinets, 2010: 188). As data collection and analyses progressed, I also became more aware of the multiple roles I played at different stages of my research. Take my collaborative Flickr project again as an example: The two key researchers, David Barton and I, were both active users of Flickr. At the time of research, we still regularly uploaded photos, added tags, and wrote captions for our photos, and we made contact with other Flickr users and left comments on their photo pages. This also means we were fairly familiar with our research context before we collected data and contacted the target participants. The participants were complete strangers to us outside the world of Flickr. All that we shared was our interest in photography and Flickr, and sometimes the languages we used. The focus on multilingual practices was partly driven by our own knowledge of an additional language besides English – I am a Chinese (Cantonese)–English bilingual and David Barton knows Spanish, which coincidentally were the two most used languages on Flickr after English (Barton and Lee, 2009). Our multilingual experiences on Flickr informed our understanding of the relationship between English and other languages online, offering some insider perspectives to our research. In other words, our position as researchers of Flickr was partly enabled by our roles as insiders and active users of Flickr. To the research participants, we were sometimes their Flickr contacts; at other times, we were their site visitors and researchers. Some of these roles were consciously and immediately reflected upon as we interacted with them, while other roles became clearer as we analyzed the data. We were especially reflexive of how our multiple roles and positions could possibly impact our research questions, our data collection processes, and how we would interpret our participants' words.

Davies and Merchant (2007: 173), in their autoethnographic study of their own academic blogging, outlined the possible roles played by new media researchers:

- Researcher as identifier of new tropes
- Researcher as insider
- Researcher as analyst
- Researcher as both subject and object
- Researcher as activist.

The first role is the researcher as the identifier of new tropes. This applies to those researchers whose aim is to discover the newness of technologies. For multilingual CMC researchers, this may involve discovering new linguistic patterns and phenomena that have never been reported on in previous research. The second role is when the researcher becomes the insider of the research site, that is, being an active user of the technology while examining other users of the technology in question. The third role is the obvious role of being the researcher-analyst, and for CMC research this can mean the analysis of texts and practices. Fourth, the researcher can be the subject and object of research at the same time. In autoethnographic research, the researcher plays the dual role of someone conducting research on something (subject), in which the researcher also becomes the research instrument (object). The last role is the researcher as activist. This often involves the researcher starting the research with a social problem. A critical discourse analysis of language ideologies on YouTube would fall into this category. In reality, the boundaries between these roles can be fuzzy, and often there are overlaps. In many cases, the researcher plays more than one of these roles. The researcher can be an insider, the research subject, and the analyst at the same time.

In my broader study on the linguistic practices and identity construction in undergraduate Facebook groups (Lee, 2014), I found myself playing complex and overlapping roles throughout the project and when writing up the study. The project was motivated by my ongoing use of the Facebook group feature in all of my undergraduate linguistics courses. As the course instructor, I regularly posted announcements, handouts, and discussion topics to the groups. Immediately, I positioned myself as an active participant and insider of my research context. Part of the study was concerned with the ways in which the medium of instruction in the physical classroom, that is, English, shaped my own and my students' language choice in the course Facebook groups. The language used in my initial postings suggests that my strong influence on my students' language choice was inevitable. For example, in one post, I greeted my students Happy Chinese New Year in Romanized Chinese: Kung Hey Fat Choy. This resulted in students greeting me back in Cantonese and mixed-language writing, which was different from the English-only academic register we used elsewhere in the group.

Another more general methodological issue arising from the academic Facebook group project is that I did not analyze the posts until the courses were over. While the courses were taught, I was already planning to research these groups. I was particularly reflexive and conscious about my form of participation as both a teacher and a researcher with an interest in carrying out further research for conference presentations, seminars, and publications, through which I share my experiences with colleagues of similar interest. Inevitably, my own posts then became part of my textual data for analysis. In addition to being reflexive, I asked myself who I was at different stages of teaching and research, including data collection and interpretation. My heavy involvement as a teacher who planned to research my students' posts gave rise to issues of ethics and validity. One possible solution, as Anderson (1998) recommends, is to analyze the data with a collaborator who is not directly involved in the research context in question. For my case, I had my research assistant, who was a Cantonese–English bilingual, interpret my own posts and my students' comments on them. This also allowed for a second opinion on the data being analyzed and interpreted.

THE MULTILINGUAL RESEARCHER

Of particular relevance to researching online multilingualism is the question of whether it is essential for the researcher to be knowledgeable in the languages used by the research participants. In my experience, linguists often complain about being asked the familiar question "How many languages do you speak?", and formal linguists have made it explicit that it is not necessary to be proficient in a language before they can study the structure of it. In sociolinguistics, however, cross-linguistic research often involves a close relationship with language users. Thus, knowing the native languages of the data and participants and their cultural backgrounds can facilitate the data collection process, to a certain extent (Li, 2000). Li (2000) adds that it is important to take into consideration the linguistic background and competence of the bilingual researcher and how these can affect the research process. In addition, it is equally important to understand the researcher's "attitude towards bilingualism" (Li, 2000: 479) as the researcher's language ideologies can significantly affect the research results. Temple (2006), however, argues that being able to speak multiple languages does not automatically enhance multilingualism research as it largely depends on what "being bilingual" means to the individuals involved in the research. She also highlights the need to investigate the researchers' roles when conducting bilingual research.

In the context of multilingual CMC research, a review of previous studies, including those in Table 8.1, indicates that most researchers either studied their own native language(s), sometimes alongside English, or a language with which they had regular contact. Some may work with collaborators who speak the languages in question. My background as a Cantonese–English bilingual growing up in Hong Kong motivated my CMC research on other Hong Kong CMC users' linguistic practices. It was also my native knowledge of Cantonese and Chinese writing that enabled me to observe closely language choice online among Hong Kong students. Having been able to interview my participants in Cantonese, our shared language, was most beneficial in establishing rapport with them. However, nativeness should not be treated as a requirement in multilingualism research; just as not all native speakers of English would understand what LOL means in a text message, knowing the participant's native language does not automatically grant the researcher access to insider knowledge of CMC writing. When I analyzed the Facebook comments written by Hong Kong students, there were Cantonese-looking expressions which I found completely incomprehensible, just because these expressions originated from a local discussion forum which I was not familiar with then. This returns to my earlier point about the importance of being an insider researcher and knowing how meanings are constructed in the research context. The internet is a heterogeneous environment with countless affinity groups, each with their own writing style, language resources, and insider knowledge. Being explicit about the researcher's linguistic background and stance toward multilingualism can definitely inform the research process.

WHERE WE ARE NOW AND THE WAY FORWARD

A decade ago, at the end of their introduction to *The Multilingual Internet*, Danet and Herring (2007: 28) outlined a few directions for future research in online multilingualism. For example, they called for "more studies that assess the actual extent and nature of CMC use in different parts of the world, rather than relying on estimates of number of Internet hosts or numbers of computers". With the advancement of technical affordances for non-Roman characters, they also questioned the future of previously unwritten languages. They asked how blogging and other forms of online self-publishing might affect people's language choices. Finally, they called for "a more diverse set of analytical methods".

From the studies discussed in this book, it is evident that most of these issues have been addressed to a large extent since 2007. First, quantitative research that aims to measure linguistic diversity has adopted a more comprehensive range of criteria, including user-based variables (Chapter 2). The rise of big data and cloud computing have certainly facilitated more accurate analyses of multilingualism. Second, advanced technology in processing fonts and scripts has not stopped speakers of lesser-written or unwritten languages from adopting creative means to represent their local languages in their digital writing. For example, Greeklish and Romanized Cantonese were still present online in 2015 (e.g. Bianchi, 2015; Lee,

2015). More languages have been examined since 2007, including minority languages. In Chapter 6, I examined the impact of the internet on minority languages and the extent to which smaller languages have been made more visible online. Two opposing findings have been noted: (1) some minority language speakers have not abandoned their native languages amidst discourses of English as an online lingua franca, and (2) some minority language users still prefer to shift to English, despite the improvement of translation technologies (Chapter 7).

The most significant change of the field over the past few years has been the sharp increase in research on social media platforms. Since the advent of Web 2.0 media, CMC is no longer confined to the synchronous–asynchronous dichotomy. Social media exhibit a "collocation of online spaces" where synchronous and asynchronous forms of CMC interaction take place in one space (Knobel and Lankshear, 2008: 265). The complex nature of social networks in newer media also gives rise to new forms of multilingual practices. Several examples of this have been discussed in this book. While the familiar practice of code-switching continues to prevail in Web 2.0 media, participants have identified new meanings and discourse functions of code-switching practices (Chapter 3). For example, the function of mixed code writing on IM is different from that of a mixed language hashtag on Instagram, partly due to their different affordances and perceived uses. Finally, as outlined in the current chapter, multilingual CMC researchers have definitely adopted a more mixed-methods approach to research. Quantitative data are complemented by ethnographic methods and data. More attention is also given to multilingualism as a resource for identity construction (Chapter 4), not just as texts. Another strand of research takes a metalinguistic angle and examines how language ideologies are represented in new media (Chapter 5). Not only has general CMC research progressed, but some essential linguistic inquiries have been revisited and new paradigms have been adopted. Crucially, as discussed in Chapter 1, the meaning of multilingualism itself has been questioned, and some researchers prefer newer terms such as *polylingualism, translingualism*, and *metrolingualism*. The question of what multilingualism means remains an open-ended one that needs constant revisiting.

While this book has provided an overview of existing multilingualism online research, future research needs to take into account some the following topics.

- *Multimodal-multilingual CMC data*: Analyzing multimodal data is not new in CMC research. However, compared to text-based CMC, studies of multimodal CMC are still limited. At the first conference on Approaches to Digital Discourse Analysis held in November 2015, the three keynote speakers, Jannis Androutsopoulos, Susan Herring, and Crispin Thurlow, coincidentally highlighted the need for studying CMC from a transmodal perspective. Rather than continuing to adopt existing linguistic theories, Herring in particular also urged CMC researchers to come up with CMC-only frameworks for analyzing the increasingly complex and multimodal nature of digital discourse (e.g. robot-mediated communication as discussed in Herring, 2015). Of course, human–computer interaction, and in fact communication in general, is *always* multimodal, although some forms of interaction may be more or less multimodal than the others. Over the years, frameworks and concepts associated with multimodality have been developed by linguists and literacy scholars (Cope and Kalantzis, 2009; Kress, 2009). Some of these have been applied in CMC research (e.g. Sindoni, 2013). There is still an urgent need to understand how multimodal data interact with concepts in multilingualism and superdiversity.

- *New structural features of CMC*: While digital language researchers have gradually moved away from describing linguistic features to understanding CMC as socially constructed discourse, it is still worthwhile to examine whether newer forms of CMC discourse give rise to new linguistic structure. For example, the formation of hashtags has certainly posed new questions for studies of morphological and semantic structures. Is a hashtag a new linguistic unit because it can be a string of words not separated by space? Are there any systematic structural patterns of forming hashtags?
- *Interaction between online and offline linguistic landscapes*: Linguistic features that are commonly considered to be specific to online communication have been increasingly commodified as they often index global identity. These features have also made their way to off-screen linguistic landscapes. Public spaces are gradually infiltrated with texts with traces of what public discourse would refer to as *textese* or *Netspeak*. This seems to be happening all around the world, as reported in Lee (2015). It is important to see whether multilingual CMC has an impact on the offline linguistic world. The online–offline divide is also beginning to break down with the rise of mobile apps that facilitate people's offline activities. It would be interesting to investigate how face-to-face multilingual discourses are mediatized and mediated by mobile apps.
- *Transmedia multilingual practices*: Much research on multilingualism has focused on one CMC platform at a time. However, people's everyday activities are increasingly dependent upon multiple CMC platforms as well as multiple devices. How do language choice and multilingual practices on one CMC platform shape those on another? How are multilingual discourses constructed through multiple devices such as the computer, smartphones, and tablets? Do people have different multilingual practices on different platforms and devices, and how does that affect their everyday linguistic activities?

From a literacy studies point of view, some of these questions can be addressed by paying close attention to people's everyday practices and collecting ethnographic data. However, more systematic analytical frameworks for researching transmodal and transmedia discourses can definitely shed light on our limited understanding of flows of linguistic resources online and offline, and across modes and media. Of course, there is no one-size-fits-all research method that is suitable for all multilingual research. Methods and theories also change according to changing technological affordances while some underlying principles remain relevant for a longer time. Multilingual CMC researchers should be encouraged to continue to experiment with a range of methods that fits in with their own research agendas, while taking the initiative in developing newer methods and approaches for future research.

Epilogue

AN AUTO-GLOSSO-TECHNOBIOGRAPHY

This book began with Carmen Lee's auto-technobiography. Similarly, I will start this epilogue by briefly relating my *auto-glosso-technobiography*, or the story of my interests in language and technology. Greek *glosso-* (language) is infixed to limit the scope of the account, since some of my interests in technology-mediated communication are orthogonal to multilingualism per se. Nonetheless, multilingualism has been a persistent thread running through my research program for the past 15 years.

I decided to study linguistics many years ago because of an interest in foreign languages, and by the time I graduated with my doctorate, I had studied a dozen or so languages. When I began researching computer-mediated communication (CMC), however, it was in my native language, English, in which I felt most confident about my linguistic intuitions. CMC was already taking place in other languages at that time, for example, in Usenet newsgroups by speakers of other languages living in the Western diaspora, but it was not until internet access spread around the globe starting in the mid-1990s that I became more than casually interested in multilingual CMC. During those years, I edited collections that included contributions on cross-cultural and non-English-language CMC, and I started encouraging my foreign students to study the uses of their native languages online.

By the end of the decade, I was growing concerned about the dominance of English in CMC research and on the internet as a whole. I expressed that concern publicly in a keynote lecture at the Cultural Attitudes towards Technology and Communication (CATaC) Conference (Herring, 2002). That lecture was the first of what would become several broad efforts to pull together existing work on, and lay out agendas for, online multilingualism research, the best known of which is the 2007 collection I co-edited with Brenda Danet, *The Multilingual Internet: Language, Culture, and Communication Online.*

Having laid out those agendas, I felt that I should do my part to contribute to fulfilling them. My own contributions to online multilingualism fall into three streams that emerged roughly in the following order, but with temporal overlap: broad overviews and agenda-setting; case studies of a single language; and cross-language comparisons. The first stream was alluded to previously. The second and third have been facilitated by the native or native-like language expertise of various co-authors, since by the turn of the millennium, I had moved from a linguistics department to a department of information science, where, unlike in linguistics, multiple co-authorship is the norm. This had the wonderful unintended benefit of expanding the scope of my CMC research beyond English. My co-authors and I have studied Thai, Lithuanian, Italian, and Bengali CMC. We have compared CMC in Lithuanian and Croatian, in English and Arabic, in English and Polish, and in English and Mandarin Chinese, as well as conducting larger cross-language comparisons, one involving more than 50 languages (discussed further later). Like multilingual CMC itself, this research has involved multiple CMC modes, including chat, blogs, text messages posted to interactive television programs, Twitter, Wikipedia, e-commerce sites, news sites, university websites, social network sites, multiplayer online games, and video.[1] Yet this body of work, albeit diverse, is only a drop in the bucket, so numerous are the languages used online now, the contexts of their use, and the possible approaches that linguists could take to study them.

TRENDS IN ONLINE MULTILINGUALISM RESEARCH

The Danet and Herring (2007) collection represented online multilingualism research broadly, with no specific focus; we sought to include the best work that was available at the time, drawing from a variety of linguistic and methodological domains.[2] It included, for example, chapters on writing systems, politeness, language choice, global linguistic diversity, and challenges posed by CMC for machine translation. The chapter by the author of the present monograph, Carmen Lee, compared the linguistic features of email and ICQ produced by Cantonese–English bilinguals in Hong Kong. That study, like most of the others in the 2007 collection, analyzed language use data from log files.

The present monograph makes a more focused and coherent contribution by privileging analyses that drill down below the surface of linguistic expression to incorporate context, in keeping with contemporary trends in micro-sociolinguistics (Androutsopoulos, 2006). By providing numerous rich, contextualized vignettes of multilingual online language use, this book complements the log-based studies that came before and advances the field of study overall. Sadly, Brenda Danet did not live to see the flowering of research on online multilingualism as described in this book. I expect that she would have appreciated it greatly, all the more so in that her own approach to research inclined toward the qualitative and the ethnographic.

Lee makes a persuasive case for a context-rich, micro-sociolinguistic approach to online multilingualism; indeed, hers is a distinctive voice that has emerged in recent years in association with this approach. Her book also addresses some broader-scope issues such as online linguistic diversity and metadiscourses about internet multilingualism. At the same time, important macro-sociolinguistic questions remain, and these should not be lost sight of in the turn toward contextualized specificity in sociolinguistic CMC research.

The macro-level questions I have in mind concern, for example, universal versus language-specific practices, language contact and spread, the status of minority and majority languages in the linguistic ecology of the internet, and multilingualism on a regional and global scale, as well as longitudinal trends in the foregoing. Comparative language research, sometimes on a large scale, is required to address these phenomena. If a study aims to go beyond simply counting languages, moreover, it may be resource intensive, requiring manual analysis of potentially massive amounts of data. Some working knowledge of multiple languages is also helpful. It follows that this line of research is often best undertaken by multilingual teams of researchers rather than by single scholars.

CROSS-LANGUAGE TRENDS IN ONLINE MULTILINGUALISM

For some years, I have been drawn to comparative studies in a quest for "big picture" understandings of online multilingualism. In what follows, I discuss three projects in which my collaborators and I analyzed four or more languages and which illustrate some of the macro-sociolinguistic issues mentioned previously. In addition, each study sheds light on a larger trend (or trends) and thus, I suggest, provides a basis for making predictions about future online language use. The first study concerns language networks across blogs; the second is a longitudinal study of language choice on university websites; and the third focuses on special language varieties associated with playful online subcultures.

Language networks on a blog-hosting site

LiveJournal.com is a blog-hosting platform and one of the first social network sites (boyd and Ellison, 2007). Created in 1999 by a 19-year-old American programmer, it was sold in 2005 to Six Apart in San Francisco. By the time of our study in 2006 (published as Herring, Paolillo, et al., 2007), the platform was international in scope and multilingual, with templates available in 32 languages. The platform's slogan was: "LiveJournal.com is a place where you can share your thoughts with the world." We wanted to know what languages were actually used on LiveJournal.com and how robust non-English language networks were. A multilingual team of eight researchers coded a random sample of LiveJournals and identified the most frequent non-English languages (English-language blogs accounted for 84% of the sample): Russian (11%), followed by Portuguese, Finnish, and Japanese (less than 1% each). We then constructed a seed sample of monolingual LiveJournals for each of the four languages and crawled two degrees of "friends" links from the seed languages to approximate social networks. The main languages used in the more than 5,000 blogs in these networks were manually identified,[3] counted, and visualized in social network graphs.

We found that the better-represented a language was on the site, the more friends bloggers in that language had, on average, and the more likely it was that friends' blogs would use the same language. The Russian network was the densest, followed by the Portuguese and the Finnish networks. The Japanese network was sparse; few Japanese blogs linked to other Japanese blogs, most linked to English blogs. In contrast, Russian blogs linked overwhelmingly to other Russian blogs. Thus, a monolingual speaker of Russian (and to a lesser extent, Portuguese) could enjoy much content on LiveJournal, whereas Finnish and Japanese speakers would not have much blog content to read or respond to unless they knew English. We concluded that LiveJournal.com was not as multilingual in practice as it presented itself as being.

Moreover, it turned out that we had inadvertently identified evidence of a larger trend. One year after our study, Six Apart sold LiveJournal.com to the Russian media company SUP, due to the popularity of the platform with Russian users. When I sampled LiveJournals randomly again in November 2010, 56% were in Russian and only 34% were in English. A similar pattern of U.S. social media platforms being co-opted by speakers of other languages can be observed for other platforms. The social network site Orkut, for example, was created in 2004 by Google in California, but after a few years, Portuguese speakers were more numerous than English speakers on the site, and in 2008, Google announced that Orkut would be managed and operated by Google Brazil. As of April 2010, 48% of Orkut's users were from Brazil, followed by India with 39% and the U.S. with only 2%. Another example is Friendster, launched in the U.S. in 2002. Its initial success in the U.S. was cut short due to dissatisfaction with the site's management, but its popularity continued to grow in Asia, especially in the Philippines, and in December 2009, Friendster was acquired by MOL Global, one of Asia's largest internet companies. Another social network site, Hi5, was created in San Francisco in 2003 and was popular in the U.S. for a while, but now it is frequented mostly by users from Central America, South America, and Thailand. Only 14% of its current users are from the U.S., mostly Hispanics and African Americans.

Based on these trends, we might venture a prediction that the U.S. social network site Facebook, which has been adopted by speakers of many other languages around the world, will someday shift its operations offshore and be replaced in the U.S. by a new, popular platform. In the meantime, Facebook remains highly popular in the U.S., while its global spread[4] continues apparently unchecked. Indeed, a number of other social network sites have been shut down due to their inability to compete with Facebook, including Orkut (in 2014) and Friendster (in 2015).

The global linguistic ecology of social network sites is a fascinating topic. Sites launched in countries other than the U.S. do not exhibit the same tendency to be taken over by speakers of other languages – why is that? It is an interesting question, moreover, how particular U.S. sites come to be popular with speakers of particular languages. Early adopters and influence leaders may be two favoring factors. (In the case of LiveJournal, several early adopters of the site were Russians studying in the U.S. who carried word of it back to Moscow; subsequently,

LiveJournal was adopted by a high-profile Russian journalist who wrote about it on his website.) Targeted marketing and economic resources, as in the case of Facebook, play a role as well. In addressing questions such as these, linguistic researchers might profitably draw from both macro-sociolinguistics and political economics (e.g. Dor, 2004).

Changing language choices on university websites

Institutions with a global audience are increasingly making versions of their websites available in different languages. Universities, in particular, are interested in marketing their institutions to foreign students and promoting international contact. Callahan and Herring (2012) investigated how often universities in different countries provide foreign-language webpages, what languages they use, and whether there has been any change over time in the languages used. More broadly, we were interested in the question of whether the internet facilitates the use of some languages (such as English) at the expense of others, and if so, which languages are favored or disfavored?

To address these questions, we manually analyzed the webpages of more than 1,100 universities in 57 countries at three points in time over a five-year period. Each website was coded for country, primary language, and secondary language(s) used. My co-author speaks Polish, Russian, and German, in addition to English, and I have some familiarity with a number of languages. That said, our broad sample included languages and writing systems that neither of us knew; in such cases, we made use of cues available on the website, such as flag icons next to the language options, and Google Translate.

Perhaps unsurprisingly, given that English is the international lingua franca of higher education, we found that 72% of the countries had English as the primary or secondary language on a majority of their university websites, and English was used as a primary or secondary language to some degree in all but one of the 57 countries. Yet the overall degree of multilingualism was also rather high: 52% of the websites had at least one additional page in another language, 16% presented information in three or more languages, and several sites also provided an option for machine translation through Google Translate. Analogously, our longitudinal analysis found that English is expanding, but the use of other languages is expanding more. Overall, the number of bilingual and multilingual sites increased between 2006 and 2011. Countries with multilingual sites included Spain (Spanish, English, Catalan/Galician), Iran (Farsi, English, Arabic), Israel (Hebrew, English, Russian), and Japan (Japanese, English, Chinese).[5]

Dor (2004) predicted that the globalization of the internet would lead to market-driven "global diglossia", with English as the language of international communication and commerce among speakers of different languages and local languages used for marketing products and services to local communities, for example. But the situation for university websites might better be characterized as "dynamic multilingualism" – multilingual in that more than two languages are involved and dynamic in that multilingualism increases as different language versions are added over time.

These findings touch on a number of broader issues. What factors contribute to university websites in some countries becoming more multilingual than in other countries? The most linguistically diverse websites in our global sample were in Australia and New Zealand and in countries in the European Union, where social and political realities favor cross-linguistic outreach. Universities in the former countries attract a broad spectrum of students from South and Southeast Asia, and countries of the European Union are participating increasingly in cooperative international scholarship programs. What do these findings say about trends in the online global linguistic ecology? Since pressures toward internationalism are currently affecting most countries on earth, we can extrapolate that the internet will become increasingly diverse. Scholars' early fears of English spreading via the internet to the detriment of other languages (cf. Herring, 2002) thus appear less and less likely to be realized; English is spreading (mainly as a lingua franca), but other languages are not being marginalized. The trend is quite the opposite.

How much can we extrapolate from university websites to other online genres, though? It seems that the purpose and audience of the websites make a difference. For example, a high level of "globalization" (translation into other languages) has been observed for websites in the electronics, telecommunication, entertainment, news, and travel industries, but less for websites in the banking, food and drink, insurance, and finance industries (Singh and Boughton, 2005). The latter provide services that are arguably oriented more toward local consumers within a single linguistic area than the former are, consistent with Dor's (2004) global diglossia model.

Finally, I noted previously that several websites in the Callahan and Herring (2012) study incorporated automated translation tools. It is not a stretch to predict that before long, visitors to any website[6] will be able to view its contents in an array of languages with the click of an icon. This technology is already available as a free Microsoft widget that can be installed on any website to translate it into more than 30 languages, including Haitian Creole, Hmong Daw, and Yucatec Mayan.[7] Widespread adoption of automated translation has the potential to alter many of linguists' basic assumptions about language choice in online communication. Choices will be tied less to the competence and identities of the users than in traditional offline communication, although perhaps the choices will be no less strategically motivated. This promises to be a richly rewarding area of study in future multilingualism research.

Cross-cultural comparison of special internet language varieties

The third study differs from the other two in that the units of analysis are much smaller (parts of words and sentences instead of webpages); nonetheless, it is a cross-linguistic study that I believe sheds light on macro-level issues.

Several years ago I became interested in *special internet language varieties* (SILVs). A SILV, as defined by Herring, Kouper, et al. (2012),[8] is a highly playful, nonstandard variant of a standard language that is more divergent from standard language norms and less comprehensible to non-initiates than typical "Netspeak". I and four collaborators analyzed the linguistic characteristics and social contexts of four SILVs, each originating in a different culture and based on a different language: Leet Speak (U.S.), Padonki (Russia), Fakatsa (Israel), and Martian Language (Taiwan and China).[9] The research team included native speakers of Russian, Hebrew, and Mandarin Chinese, as well as a member with a background in computer science who was familiar with U.S. hacker culture. We asked: How similar are SILVs across cultures? If similarities are present, what accounts for them (contact across varieties, common generative linguistic principles, common social factors)?

Our data source was texts from the public internet contexts in which each SILV was used most often – bulletin boards, game chat logs, blogs, web forum posts and comments, and so on, depending on the SILV. We categorized and compared the features of each language subsample at multiple linguistic levels: typographic, orthographic, phonological, lexical, morphological, syntactic, pragmatic, and rhetorical. With the exception of Padonki, the features most involved in SILV production were mostly typographic and orthographic. These types co-occurred so often that we coined a portmanteau term, *typthographic*, to refer to them.

Our comparison revealed a number of similarities across the SILVs but little evidence of cross-variety contact. Rather, the similarities appeared to derive from the exploitation of common principles, such as substituting certain characters for other characters based on graphical or sound resemblance, combined with the metalinguistic awareness fostered by persistent textual CMC, which lends itself to language play. These principles might be considered candidates for typthographic universals; additional evidence for them can be found in several of the chapters in Danet and Herring (2007). Moreover, all four varieties arose in online subcultures where there was frequent in-group interaction, and they seemed to be motivated by a desire on the part of their users to make their writing distinctive (e.g., obscure, humorous, decorative). That is, SILV use signaled in-group identification.

Differences among the varieties appeared to be conditioned largely by the resources made available by the writing system of each base language (in the case of typthography) and the sociocultural context in which each SILV arose. Leet users were primarily young and male, and their subcultures valued hacking and computer skills. Padonki users were educated males in their late 20s and 30s with antisocial, contrarian values. Fakatsa users were preteen girls, and the subculture was focused on cuteness, femininity, and perfection.[10] The use of Martian Language was also associated with female more than male users and expressed ironic cuteness, especially in regard to romantic love. These values were reflected in language use in the SILVs at all levels, for example in the affixation of nonsense syllables to Hebrew words in Fakatsa to represent a cute, childish style of speech, or the proliferation of creative variants of Russian profanities in Padonki.

In addition to suggesting potential universals of creative online typthography, the SILV study sheds light on how linguistic innovations arise and spread. Some features of the four SILVs have made their way into general internet language usage; for example, Leet n00b ("newbie") and w00t (an expression of excitement) have become part of English Netspeak. Some SILV expressions have also percolated up into offline use, for example on billboards, T-shirts, record albums, and book covers, where their usage is mostly tongue-in-cheek to index hipness and internet savvy. By the time we observed such uses, however, the SILVs themselves had already become passé, much like youth slang when it is adopted by adults. As with slang and other language fads, SILVs are transitional phenomena; at the time we collected our data, most of the SILVs had already peaked in usage.

We might predict that new SILVs will arise to take the place of old ones in these languages, and that SILVs will be found in other languages, given that online subcultures arise in many linguistic contexts and that CMC promotes metalinguistic reflection and language play. There are also other SILV-like varieties, such as offline special language varieties that have migrated online (e.g. German rapper language; Androutsopoulos, 2007) and graphics-based forms of subcultural communication. Indeed, SILVs today are more likely to incorporate multimodal elements than not. The phenomena of LOLspeak – for example "I can has cheezburger?" superimposed over a photo of a cat, and Doge Speak, such as "many happy" and "very love" superimposed over images of a shiba inu dog –[11] are arguably graphical SILVs. An understanding of text-based SILVs can provide a useful comparative lens through which to identify and study these phenomena. More generally, how special language varieties online arise, spread, and die out; what, if anything, is universal versus what is particular to each variety; and what role the internet plays in these processes constitute questions that should be of interest to students of internet multilingualism.

The three studies described in the previous sections illustrate very different methodological approaches to cross-cultural comparison: social network analysis, content analysis, and close qualitative analysis of linguistic features combined with sociocultural interpretation. Through participating in these studies, I have come to understand multilingualism online in new and, I believe, ultimately complementary ways, despite how disparate the phenomena may appear. All are aspects of the "big picture" of online language use, arrived at by comparing across multiple cultures or subcultures in order to arrive at more general insights, which in turn may serve to predict future trends.

CLOSING THOUGHTS

With the publication of *Multilingualism Online*, the field of internet multilingualism research has attained a new level of maturity. In the future, we can expect to see increased specialization and development of subareas within this field. Areas that could give rise to future book-length works include CMC in lesser-studied and minority languages; ethnographic studies of language use and language choice; cross-cultural communication and nonnative language use; multilingualism in contemporary social media; and macro-level studies of degrees of, and trends in, internet multilingualism.

Meanwhile, new technologies raise new questions and can be expected to lead to new areas of inquiry. What will the effects of widespread online machine translation be, for example, on cross-cultural communication? When access to other languages is a click away, for what purposes will it be used, and to what effect? It seems to me that the future holds some rather exciting prospects.

Susan C. Herring

NOTES

1 A list of my publications, with links, can be found at: http://info.ils.indiana.edu/~herring/pubs.html.
2 An earlier version of the collection, containing a subset of the articles in Danet and Herring (2007), was published in 2003 as a special issue of the *Journal of Computer-Mediated Communication.*
3 Our research team included a native speaker of Russian and a speaker of Portuguese, and I have studied some Japanese. Finnish was identified, after some online research, through the use of characteristic words and letter sequences, especially those involving diacritics.
4 As of January 2016, it was the most popular social network site in 129 out of 137 countries worldwide (http://vincos.it/world-map-of-social-networks/, accessed March 21, 2016).
5 This is a simplified summary of a complex global picture. For further details, see Callahan and Herring (2012).
6 Any website that uses a standard language variety, that is. Nonstandard and creative language use may continue to pose challenges for machine translation (see Climent et al. 2007).
7 www.microsoft.com/Web/solutions/mstranslator.aspx, accessed March 20, 2016.
8 This research is unpublished, but a video of a 2012 presentation is available that includes further information and examples of language use in SILVs.
9 The names of the varieties are endonyms, that is, those used in the subcultures themselves. The name Leet (also written as 1337, 133+, etc.) is derived from the word *elite.* Padonki is a deliberate misspelling of the Russian word for scoundrels or scumbags. Padonki is also known as Padonkaffsky jargon. Fakatsa is an acronym in Hebrew that translates roughly as "a shallow, stupid, noisy girl." Martian Language is so named because it is thought to be complex and unreadable.
10 See Vaisman (2014) for a more in-depth analysis of Fakatsa.
11 For more on these varieties, see Gawne and Vaughan (2011) and McCulloch (2014).

REFERENCES

Androutsopoulos, J. (2006) 'Introduction: Sociolinguistics and computer-mediated communication', *Journal of Sociolinguistics,* 10(4): 419–438.

Androutsopoulos, J. (2007) 'Style online: Doing hip-hop on the German-speaking Web', in P. Auer (ed) *Style and social identities: Alternative approaches to linguistic heterogeneity*, Berlin/New York: de Gruyter.

boyd, d. and Ellison, N. (2007) 'Social network sites: Definition, history, and scholarship', *Journal of Computer Mediated Communication,* 13(1): 210–230.

Callahan, E. and Herring, S.C. (2012) 'Language choice on university websites: Longitudinal trends', *International Journal of Communication,* 6: 322–355.

Climent, S., More, J., Oliver, A., Salvatierra, M., Sanchez, I., Taule, M. and Vallmanya, L. (2003) 'Bilingual newsgroups in Catalonia: A challenge for machine translation', *Journal of Computer-Mediated Communication,* 9(November). http://onlinelibrary.wiley.com/doi/10.1111/j.1083-6101.2003.tb00360.x/full

Danet, B. and Herring, S.C. (eds) (2007) *The multilingual Internet: Language, culture, and communication online*, New York: Oxford University Press.

Dor, D. (2004) 'From Englishization to imposed multilingualism: Globalization, the Internet, and the political economy of the linguistic code', *Public Culture,* 16(1): 97–118.

Gawne, L. and Vaughan, J. (2011) 'I can haz language play: The construction of language and identity in LOLspeak', *Proceedings of the 42nd Australian Linguistic Society Conference.* https://digitalcollections.anu.edu.au/bitstream/1885/9398/5/Gawne_ICanHaz2012.pdf

Herring, S.C. (2002) *The language of the Internet: English dominance or heteroglossia?* Keynote talk, Conference on Cultural Attitudes towards Technology and Communication (CATaC), Université de Montréal, July 13.

Herring, S.C., Kouper, I., Kutz, D., Vaisman and Zhang, G. (2012) Linguistic creativity online: A cross-cultural study of special Internet language varieties. Pragmatics Festival, Indiana University, Bloomington. Video presentation available at http://vimeo.com/40743805

Herring, S.C., Paolillo, J.C., Ramos Vielba, I., Kouper, I., Wright, E., Stoerger, S., Scheidt, L.A. and Clark, B. (2007) 'Language networks on LiveJournal', *Proceedings of the Fortieth Hawaii International Conference on System Sciences*. Los Alamitos, CA: IEEE Press. http://ella.slis.indiana.edu/~herring/hicss07.pdf

McCulloch, G. (2014) 'A linguist explains the grammar of Doge. Wow', *The Toast*, February 6. http://the-toast.net/2014/02/06/linguist-explains-grammar-doge-wow/

Singh, N. and Boughton, P.D. (2005) 'Measuring web site globalization: A cross-sectional country and industry level analysis', *Journal of Website Promotion*, 1(3): 3–20.

Vaisman, C. (2014) 'Beautiful script, cute spelling and glamorous words: Doing girlhood through language playfulness on Israeli blogs', *Language & Communication*, 34: 69–80.

APPENDIX

Tony's technolinguistic biography

Tony was a third-year undergraduate student majoring in English language education at the Chinese University of Hong Kong. In other words, he was training to be a teacher of English for speakers of other languages. He reported to have about 10 years of computing experience and was a regular user of email, IM, Facebook, forums, and blogs.

Tracing his life history, in the beginning, it was his elder brother who introduced him to the internet world – playing games online with strangers from other parts of the world. That was when he was about 9 or 10 years old. He said he was too little to have his own computer account, so his elder brother helped him sign up at an online game site. He recalled that all the instructions and interfaces of the games that he used to play were written in English only. The language was too complicated for him then, he said, but he could work out how the games were played by trial and error. Still, Tony wanted to interact with overseas gamers who he had never met. So he decided to use simple English phrases such as "good" or "good game" to socialize with other gamers.

Getting to his senior high school years, he started chatting with friends quite frequently on IM (MSN Messenger). That was also the time when Tony started to take up the social affordances of Chinese messages, as they sounded friendlier and could more accurately represent himself. He explained:

> I can think faster in Chinese so I can form a sentence easily. . . . If I write in English (online) I have to check my grammar.
>
> When chatting in MSN, I use Chinese most of the time. . . . Most of my friends speak Cantonese. . . . We know each other very well. . . . Communicating in

> Cantonese with them is more accurate than in English. . . . I can say what I want to say in Cantonese . . . and there are "fashionable" words in Cantonese that can't be captured in English. . . .

Different senses of the Chinese language were mentioned in Tony's technolinguistic biography. From time to time, he would visit and post in Mainland Chinese–based online discussion forums, on which most participants interacted in Mandarin Chinese, who were not used to mixing languages in their everyday communication as Hongkongers would. Tony however explained:

> I am a Chinese too. . . . Chinese is my mother tongue. . . . When I participate in Chinese-based forums, most people use Chinese. . . . If you switch to English, that shows you are a Hong Kong person. . . . I am not saying that I am special . . . but using English may offend the mainland Chinese forum participants.

Here, Chinese is seen as the language that validates his Chinese identity. But at the same time, he was well aware of his Hong Kong identity, too, with a rather different historical and political background, thus a different range of linguistic resources to draw upon. This is how Tony projects his perceived national identity online, which is only one of the many aspects of identity that he performed online through language choice.

Tony's technolinguistic life also revolved around his student identity. In the interviews, he often referred to his major subject and how his language choice was affected by his identity as an English language major. Tony gave me this anecdote which made reference to his involvement in a student society and how that affects his attitudes toward online language choice:

> I used to be the chairperson of the EngEd [English Education] society. I once wrote a formal email to my cabinet members. One of them then wrote back a long email in English. I was so angry. The official language of the student society was Chinese, while English was just a supplementary language. I am sure he knew the policy and he also knew that I understood Chinese . . . why write back in English? I even had to look words up in a dictionary! I was offended!

From time to time, Tony explicitly restated his preferred language online:

> Although my major subject is English language education, my friends know that I prefer Chinese (when communicating online).

However, it seemed that this preference for Chinese was restricted mainly to private and interpersonal communication. Tony, as a preservice teacher, kept in touch with the students in his teaching practice school on Facebook. For this particular audience, Tony signed up for a new Facebook account where he calls himself "Teaching Tony".

> I started this teacher Facebook account towards the end of my teaching practice. I was worried that if my students discovered my "real" Facebook account, I had to reshape my identity for them.

When asked what he meant by *reshaping* his identity, Tony explained:

> When interacting with students online, I always use English. Otherwise, I can't establish my image [as an English language teacher].

Compared to his other Facebook account, Tony posted less regularly on this teacher site. He also took on a more academic discourse style and he would only share posts that are of interest to his students, such as a link to an online English dictionary, posts about the progress of his teaching and grading work, and so on. He wrote almost all the posts in English, because the target readers were students and his colleagues from the school where he did his teaching practice. He said it would have been inappropriate to use Chinese there because as an English teacher, he had to stick to this medium of instruction in order to encourage students to write to him in English too.

Doing his best to maintain his image as a school teacher, Tony was very conscious of privacy issues and his level of self-exposure on his teacher Facebook wall. On his teacher Facebook, he tended to shift to a more formal, noninterpersonal style of writing:

> When I leave a comment on my colleague's Facebook page, I know that some of his students may be able to see it . . . so I pretend to sound serious and formal . . . so that our students would think that we are talking about something constructive. . . . I take time to polish and edit [my comments].

As of early 2012, Tony had graduated and was teaching English at a local secondary school. He was still working hard to maintain and juggle between these different aspects of his public and private senses of self through linguistic means. His teacher Facebook account still had fewer, but more polished, English posts addressing a group of students; his other account remained an expressive and affective site where he interacted with his real friends in his nonteaching life.

EPILOGUE

Tony's case was revisited in October 2015. At that time, Tony had quit teaching. He went on a "working holiday" in Ireland in 2014. Because he had never been lived overseas before, this experience had strong impact on his life, as well as his language choice online. Chinese, to Tony, was still his primary language online. He said: "Nationality is not the same as race – Chinese is still the language of my race, whether you like it or not." He was always fascinated by Irish culture. Whenever he posted in English, he was posting something about Irish culture. Another life event that also shaped his Facebook practices was the 2014 Hong Kong umbrella movement. During the movement, he regularly posted images he took from the protest sites. He had posted more images than words since then, because he could no longer find the right words to describe his feelings about various recent social events. To date, his "Teaching Tony" Facebook site has remained inactive.

Bibliography

Abbott, C. (1998) 'Making connections: Young people and the Internet', in J. Sefton-Green (ed) *Digital diversions: Youth culture in the age of multimedia*, London and New York: Routledge.

Anastasiou, D. and Gupta, R. (2011) 'Comparison of crowdsourcing translation with machine translation', *Journal of Information Science*, *37*(6): 637–59.

Anazawa, R., Ishikawa, H., Park, M. J. and Kiuchi, T. (2012) 'Preliminary study of online machine translation use of nursing literature: Quality evaluation and perceived usability', *BMC Research Notes*, *5*(1): 635–41.

Anderson, G. L. (1998) 'Toward authentic participation: Deconstructing the discourses of participatory reforms in education', *American Educational Research Journal*, *35*(4): 571–603.

Androutsopoulos, J. (2006) 'Multilingualism, diaspora, and the Internet: Codes and identities on German-based diaspora websites', *Journal of Sociolinguistics*, *10*(4): 520–47.

Androutsopoulos, J. (2007) 'Language choice and code-switching in German-based diasporic web forums', in B. Danet and S. C. Herring (eds) *The multilingual Internet: Language, culture, and communication online*, New York and Oxford: Cambridge University Press.

Androutsopoulos, J. (2008) Potentials and limitations of discourse-centered online ethnography, *Language@Internet*, 5, article 8, retrieved from www.languageatinternet.de/articles/2008 (accessed 1 May 2012).

Androutsopoulos, J. (2010) 'Localizing the global on the participatory web', in N. Coupland (ed) *The handbook of language and globalization*, Malden: Blackwell Publishing.

Androutsopoulos, J. (2012) 'English "on top": Discourse functions of English resources in the German mediascape', *Sociolinguistic Studies*, *6*(2): 209–38.

Androutsopoulos, J. (2013a) 'Networked multilingualism: Some language practices on Facebook and their implications', *International Journal of Bilingualism*, retrieved from http://ijb.sagepub.com/content/early/2013/06/07/1367006913489198.full.pdf (accessed 21 January 2016).

Androutsopoulos, J. (2013b) 'Code-switching in computer-mediated communication', in S. C. Herring, D. Stein and T. Virtanen (eds) *Pragmatics of computer-mediated communication*, Berlin: Mouton de Gruyter.

Androutsopoulos, J. (2013c) 'Participatory culture and metalinguistic discourse: Performing and negotiating German dialects on YouTube', in D. Tannen and A. M. Trester (eds) *Discourse 2: Language and new media*, Washington, DC: Georgetown University Press.

Androutsopoulos, J. (2015) 'Negotiating authenticities in mediatized times', *Discourse, Context & Media*, 8: 74–7.

Androutsopoulos, J. and Juffermans, K. (2014) 'Digital language practices in superdiversity: Introduction', *Discourse, Context & Media*, 4–5: 1–6.

Androutsopoulos, J. and Schmidt, G. (2002) 'SMS-Kommunikation: Ethnografische Gattungsanalyse am Beispiel einer Kleingruppe', *Zeitschrift für angewandte Linguistik*, 36: 49–80.

Anis, J. (ed) (1999) *Internet communication et langue française*, Paris: Hermès science publications.

Aranda, L. V. (2007) *Handbook of Spanish-English translation*, Lanham, Maryland: University Press of America.

Auer, P. (ed) (1998) *Code-switching in conversation: Language, interaction and identity*, London: Routledge.

Auer, P. (1999) 'From codeswitching via language mixing to fused lects: Toward a dynamic typology of bilingual speech', *International Journal of Bilingualism*, 3(4): 309–32.

Azuma, J. and Ebner, M. (2007) 'A stylistic analysis of graphic emoticons: Can they be candidates for a universal visual language of the future?', in J. Luca and E. R. Weippl (eds) *Proceedings of the World Conference on Educational Multimedia, Hypermedia and Telecommunications 2008*(1): 972–979. Waynesville, NC: Association for the Advancement of Computing in Education.

Babel. (1997) Web languages hit parade, retrieved from http:babel.alis.com.8080palmares.html (accessed 5 April 2000).

Bali, K., Sharma, J., Choudhury, M. and Vyas, Y. (2014) ' "I am borrowing ya mixing?" An analysis of English-Hindi code mixing in Facebook', in *Proceedings of the 2014 Conference on Empirical Methods in Natural Language Processing*: 116–126. Doha, Qatar: The Association for Computational Linguistics.

Barton, D. (1994) *Literacy: An introduction to the ecology of written language*, 1st edn, Oxford: Blackwell.

Barton, D. (2012) 'Participation, deliberate learning and discourses of learning online', *Language and Education*, 26(2): 139–50.

Barton, D. (2015a) 'Tagging on Flickr as a social practice', in R. H. Jones, A. C. Christoph and A. Hafner (eds) *Discourse and digital practices: Doing discourse analysis in the digital age*, New York: Routledge.

Barton, D. (2015b) What annoys you about technology at work? Affect and academics' discursive construction of digital writing practices, paper presented at the 1st Approaches to Digital Discourse Analysis conference, Valencia, Spain, 18–20 November 2015.

Barton, D. and Hamilton, M. (1998) *Local Literacies: Reading and writing in one community*, London: Routledge.

Barton, D. and Lee, C. (2009) 'English and glocal identities on Web 2.0: The case of Flickr.com', in K. K. Tam (ed) *Englishization in Asia*, Hong Kong: Open University of Hong Kong Press.

Barton, D. and Lee, C. (2012) 'Redefining vernacular literacies in the age of web 2.0', *Applied Linguistics*, 33(3): 282–98.

Barton, D. and Lee, C. (2013) *Language online: Investigating digital texts and practices*, London and New York: Routledge.

Bechar-Israeli H. (1995) 'From <Bonehead> to <cLonehEad>: Nicknames, play and identity on Internet relay chat', *Journal of Computer-Mediated Communication*, 1(2), retrieved from http://jcmc.indiana.edu/vol1/issue2/bechar.html (accessed 1 May 2012).

Belling, L. and De Bres, J. (2014) 'Digital superdiversity in Luxembourg: The role of Luxembourgish in a multilingual Facebook group', *Discourse, Context & Media*, 4: 74–86.

Benor, S. B. (2011) 'Mensch, bentsh, and balagan: Variation in the American Jewish linguistic repertoire', *Language & Communication*, 31(2): 141–154.

Benson, P. (2010) 'Funny teacher saying foul language: New literacies in a second language', paper presented at the 17th International Conference on Learning, Hong Kong Institute of Education, 6–9 July 2010.

Benson, P. (2013) *Teaching and researching: Autonomy in language learning*, London and New York: Routledge.

Benson, P. (2015) 'YouTube as text', in R. H. Jones, A. C. Christoph and A. Hafner (eds) *Discourse and digital practices: Doing discourse analysis in the digital age*, New York: Routledge.

Benson, P. and Chan, N. (2010) 'TESOL after YouTube: Fansubbing and informal language learning', *Taiwan Journal of TESOL*, 7(2): 1–23.

Benson, P. and Chan, N. (2011) 'TESOL after YouTube: Fansubbing and informal language learning', *Taiwan Journal of TESOL*, 7(2): 1–23.

Benson, P. and Chik, A. (2010) 'New literacies and autonomy in foreign language learning', in M. J. Luzón, M. N. Ruiz and M. L. Villanueva (eds) *Digital genres, new literacies, and autonomy in language learning*, Newcastle upon Tyne: Cambridge Scholars Publishing.

Bianchi, R. (2015) '3arabizi, Greeklish, and SMSki: The hybrid making of language in the age of the internet and mobile technology', *Tasmeem*, 1: 1–4.

Bjørnsson, J. A. (2010) *Egyptian Romanized Arabic: A study of selected features from communication among Egyptian youth on Facebook*. MA Thesis, University of Oslo.

Black, R. (2009) 'English language learners, fan communities, and twenty-first century skills', *Journal of Adolescent and Adult Literacy*, 52(8): 688–697.

Blackledge, A. (2005) *Discourse and power in a multilingual world*, Amsterdam and Philadelphia: John Benjamins.

Blackledge, A. and Creese, A. (2013) 'Heteroglossia in English complementary schools', in Gogolin, J. and J. Duarte (eds) *Linguistic super-diversity in urban areas*, Amsterdam: John Benjamins.

Blommaert, J. (2011) 'Supervernaculars and their dialects', *Working Papers in Urban Language & Literacies*, paper 81, retrieved from www.kcl.ac.uk/innovation/groups/ldc/publications/working papers/WP81.pdf (accessed 1 May 2012).

Blommaert, J. and Rampton, B. (2011) 'Language and superdiversity: A position paper', *Working Papers in Urban Languages and Literacies*, paper 70, retrieved from www.kcl.ac.uk/projects/ldc/LDCPublications/workingpapers/70.pdf (accessed 1 May 2012).

Bolander, B. and Locher, M. A. (2014) 'Doing sociolinguistic research on computer-mediated data: A review of four methodological issues', *Discourse, Context & Media*, 3: 14–26.

Bolton, K. (2000) 'The sociolinguistics of Hong Kong and the space for Hong Kong English', *World Englishes*, 19(3): 265–285.

Bourdieu, P. (1990) *The logic of practice*, Cambridge: Polity.

boyd, d. (2008) 'Facebook\'s privacy trainwreck', *Convergence: The International Journal of Research into New Media Technologies*, 14(1): 13–20.

Brabham, D. C. (2012) 'The myth of amateur crowds: A critical discourse analysis of crowdsourcing coverage', *Information, Communication & Society*, 15(3): 394–410.

Bruns, A. and Burgess, J. (2014) Twitter hashtags from ad hoc to calculated publics, retrieved from http://snurb.info/files/2015/Twitter%20Hashtags%20from%20Ad%20Hoc%20to%20Calculated%20Publics.pdf (accessed 23 January 2016).

Bruns, A. and Highfield, T. (2015) 'From news blogs to news on Twitter: Gatewatching and collaborative news curation', in S. Coleman and D. Freelon (eds) *Handbook of digital politics*, Northampton: Edward Elgar Publishing.

Busch, B., Aziza, J. and Angelika, T. (2006) *Language biographies for multilingual learning*, Cape Town: Occasional Papers 24.

Callahan, E. and Herring, S. C. (2012) 'Language choice on university websites: Longitudinal trends', *International Journal of Communication*, 6: 322–355.

Canagarajah, S. (2011) 'Codemeshing in academic writing: Identifying teachable strategies of translanguaging', *The Modern Language Journal*, 95(3): 401–417.

Canagarajah, S. (2013) *Translingual practice: Global Englishes and cosmopolitan relations*, London and New York: Routledge.

Castells, M. (1999) 'Grassrooting the space of flows', *Urban Geography*, 20(4): 294–302.

Castells, M. (ed) (2006) *The network society: From knowledge to policy*, Washington, DC: Centre for Transatlantic Relations.

Castells, M. (2010) *End of millennium: The information age: Economy, society, and culture*, Chichester, UK: John Wiley and Sons.

Cenoz, J. and Gorter, D. (2011) 'Focus on multilingualism: A study of trilingual writing', *The Modern Language Journal*, 95(3): 356–369.

Chang, H. (2009) 'Online social support: Which posts were answered?', *Journal of Contemporary Eastern Asia*, *8*(1): 31–46.

Chau, C. Y. (2014) *Facebook group for Hong Kong undergraduates: Linguistic practices and construction of identities*. Unpublished MPhil Thesis. The Chinese University of Hong Kong, Hong Kong, China.

Chen, K. H. Y. (2005) 'The social distinctiveness of two code-mixing styles in Hong Kong', in J. Cohen, K. T. McAlister, K. Rolstad, & J. MacSwan (eds) *Proceedings of the 4th International Symposium on Bilingualism*. Somerville, MA: Cascadilla Press.

Chen, S. Y. R. (2007) 'Code-switching between English and Mandarin Chinese on postings in the college-affiliated bulletin board system in Taiwan: A functional approach', in C. Gabrielatos, R. Slessor and J. W. Unger (eds) *Papers from the Lancaster University Postgraduate Conference in Linguistics & Language Teaching, Volume 1*.

Chik, A. (2015) 'Recreational language learning and digital practices', in R. H. Jones, A. C. Christoph and A. Hafner (eds) *Discourse and digital practices: Doing discourse analysis in the digital age*, New York: Routledge.

Ching, C. C. and Vigdor, L. (2005) 'Technobiographies: Perspectives from education and the arts', paper presented at the First International Congress of Qualitative Inquiry, University of Illinois, 5–7 May 2005.

Chun, E. and Walters, K. (2011) 'Orienting to Arab Orientalisms: Language, race, and humor in a YouTube video', in C. Thurlow and K. Mroczek (eds) *Digital discourse: Language in the new media*, New York and London: Oxford University Press.

Climent, S., Moré, J., Oliver, A., Salvatierra, M., Sànchez, I. and Taulé, M. (2007) 'Enhancing the status of Catalan versus Spanish in online academic forums: Obstacles to machine translation', in B. Danet and S. C. Herring (eds) *The multilingual Internet: Language, culture, and communication online*, Oxford and New York: Oxford University Press.

Climent, S., Moré, J., Oliver, A., Salvatierra, M., Sànchez, I., Taulé, M. and Vallmanya, L. (2003) 'Bilingual newsgroups in Catalonia: A challenge for machine translation', *Journal of Computer-Mediated Communication*, 9(1), retrieved from http://onlinelibrary.wiley.com/doi/10.1111/j.1083–6101.2003.tb00360.x/full (accessed 12 January 2016).

Cope, B. and Kalantzis, M. (2009) '"Multiliteracies": New literacies, new learning', *Pedagogies: An international journal*, 4(3): 164-195.

Crandall, J. (2007) Showing, retrieved from http://jordancrandall.com/showing/index.html (accessed 1 May 2012).

Creese, A. and Blackledge, A. (2010) 'Towards a sociolinguistics of superdiversity', *Zeitschrift für Erziehungswissenschaft*, *13*(4): 549–572.

Crystal, D. (1997) *English as a global language*, Cambridge: Cambridge University Press.

Crystal, D. (2001) *Language and the internet*, Cambridge: Cambridge University Press.

Crystal, D. (2006) *Language and the internet*, 2nd edn, Cambridge: Cambridge University Press.

Crystal, D. (2011) *Internet linguistics: A student guide*, London: Routledge.

Cunliffe, D. (2007) 'Minority languages and the internet: New threats, new opportunities', in M. J. Cormack and N. Hourigan (eds) *Minority language media: Concepts, critiques and case studies*, Clevedon, UK: Multilingual Matters.

Cunliffe, D. and Harries, R. (2005) 'Promoting minority-language use in a bilingual online community', *New Review of Hypermedia and Multimedia*, *11*(2): 157–179.

Cunliffe, D. and Herring, S. C. (2005) 'Introduction to minority languages, multimedia and the web', *New Review of Hypermedia and Multimedia*, *11*(2): 131–137.

Dajun, Z. and Yun, W. (2015, January) 'Corpus-based machine translation: Its current development and perspectives', *International Forum of Teaching and Studies*, *11*(1/2): 90.

Danet, B. (2001) *Cyberpl@ y: Communicating online*, Oxford, UK: Berg.

Danet, B. (2010) 'Computer-mediated English', in J. Maybin and J. Swann (eds) *The Routledge companion to English language studies*, London and New York: Routledge.

Danet, B. and Herring, S. C. (eds) (2007) *The multilingual internet: Language, culture, and communication online*, New York and Oxford: Oxford University Press.

Daoudi, A. (2011) 'Globalisation and e-Arabic: The emergence of a new language at the literal and figurative levels', *Studies in Slavic and General Linguistics*, *38*: 61–76.

Das, A. and Herring, S. C. (2016) 'Greetings and interpersonal closeness: The case of Bengalis on Orkut', *Language & Communication*, *47*: 53–65.

Davies, C. A. (1999) *Reflexive ethnography*, London and New York: Routledge.

Davies, J. (2004) 'Negotiating femininities on-Line', *Gender and Education*, *16*(1): 35–49.

Davies, J. (2007) 'Display, identity and the everyday: Self-presentation through online image sharing', *Discourse*, *28*(4): 549–64.

Davies, J. and Merchant, G. (2007) 'Looking from the inside out: Academic blogging as new literacy', in M. Knobel and C. Lankshear (eds) *A new literacies sampler*, New York: Peter Lang.

Davies, J. and Merchant, G. (2009) *Web 2.0 for schools*, New York: Peter Lang.

Derks, D., Bos, A. E. R. and von Grumbkow, J. (2007) 'Emoticons and social interactions on the internet: The importance of social context', *Computers in Human Behavior*, *23*: 842–849.

Deumert, A. (2014) *Sociolinguistics and mobile communication*, Edinburgh, UK: Edinburgh University Press.

Deumert, A. and Masinyana, S. (2008) 'Mobile language choices – The use of English and isiXhosa in text messages (SMS) Evidence from a bilingual South African sample', *English World-Wide*, *29*(2): 117–147.

Dewey, M. (2007). 'English as a lingua franca and globalization: an interconnected perspective,' *International Journal of Applied Linguistics*, *17*: 332–354.

Diakopoulos, N. and Naaman, M. (2011) 'Towards quality discourse in online news comments', *Proceedings of the ACM 2011 Conference on Computer Supported Cooperative Work*, Hangzhou, China, 19–23 March 2011.

DiMicco, J. M. and Millen, D. R. (2007) 'Identity management: Multiple presentations of self in facebook', *Proceedings of the 2007 international ACM conference on Supporting group work*, Sanibel Island, FL, USA, 4–7 November 2007.

Dor, D. (2004) 'From Englishization to imposed multilingualism: Globalization, the Internet, and the political economy of the linguistic code', *Public Culture*, *16*(1): 97–118.

Dresner, E. and Herring, S. C. (2010) 'Functions of the nonverbal in CMC: Emoticons and illocutionary force', *Communication Theory*, *20*(3): 249–268.

Durham, M. (2007) 'Language choice on a Swiss mailing list', in B. Danet and S. C. Herring (eds) *The multilingual internet: Language, culture, and communication online*, Oxford, UK: Oxford University Press.

Dyson, L. E., Hendriks, M. A. N. and Grant, S. (2007) *Information technology and indigenous people*, Hershey, PA: Information Science Publishing.

Eisenlohr, P. (2004) 'Language revitalization and new technologies: Cultures of electronic mediation and the refiguring of communities', *Annual Review of Anthropology*, *33*: 21–45.

Ensslin, A. (2013) '"What an un-wiki way of doing things": Wikipedia's multilingual policy and metalinguistic practice', in H. Kelly-Holmes and T. M. Milani (eds) *Thematising multilingualism in the media*, Amsterdam and Philadelphia: John Benjiamins Publishing Company.

Ethnologue. (2015) Languages of the world: Summary by language size, retrieved from www.ethnologue.com/statistics/size (accessed 23 June 2015).

European Charter for Regional or Minority Languages. (1992) Council of Europe, retrieved from https://rm.coe.int/CoERMPublicCommonSearchServices/DisplayDCTMContent?documentId=090000168007bf4b (accessed 26 May 2016).

Fernandez, L. (2001) 'Patterns of linguistic discrimination in Internet discussion forums', *Mercator Media Forum*, *5*(1): 22–41.

Ferrara, K., Brunner, H. and Whittemore, G. (1991) 'Interactive written discourse as an emergent register', *Written Communication*, *8*(1): 8–34.

Fishman, J. (1998) 'The new linguistic order', *Foreign Policy*, *113*: 26–40.

Fishman, J. A. (ed) (2001) *Can threatened languages be saved?: Reversing language shift, revisited: A 21st century perspective*, Clevedon, UK: Multilingual Matters.

Flammia, M. and Saunders, C. (2007) 'Language as power on the Internet', *Journal of the American Society for Information Science and Technology*, *58*(12): 1899–1903.

Fleming, A. and Debski, R. (2007) 'The use of Irish in networked communications: A study of schoolchildren in different language settings', *Journal of Multilingual and Multicultural Development*, *28*(2): 85–101.

Flickr. (n.d.) Retrieved from www.flickr.com.

Fung, L. and Carter, R. (2007) 'New varieties, new creativities: ICQ and English-Cantonese e-discourse', *Language and Literature*, *16*(4): 345–366.

Galla, C. (2009) 'Indigenous language revitalization and technology from traditional to contemporary domains', in J. A. Reyhner and L. Lockard (eds) *Indigenous language revitalization: Encouragement, guidance & lessons learned*, Flagstaff: Northern Arizona University.

Gao, X. (2012) '"Cantonese is not a dialect": Chinese netizens' defence of Cantonese as a regional lingua franca', *Journal of Multilingual and Multicultural Development*, *33*(5): 449–464.

García, O. (2009) 'Education, multilingualism and translanguaging in the 21st century', in T. Skutnabb-Kangas, R. Phillipson, A. K. Mohanty and M. Panda (eds) *Social justice through multilingual education*, Bristol, UK: Multilingual Matters.

García, O. and Li, W. (2014) *Translanguaging: Language, bilingualism and education*, London, UK: Palgrave Macmillan Pivot.

Gee, J. P. (2004) *Situated language and learning*, London: Routledge.

Gee, J. P. and Hayes, E. (2011) *Language and learning in the digital age*, London and New York: Routledge.

Georgakopoulou, A. (1997) 'Self-presentation and interactional alliances in e-mail discourse: The style-and code-switches of Greek messages', *International Journal of Applied Linguistics*, *7*(2): 141–164.

Georgakopoulou, A. (2004) 'To tell or not to tell?: Email stories between on-and off-line interactions', *Language@Internet*, *1*: 1–38.

Georgakopoulou, A. (2011) "On for drinkies?': E-mail cues of participant alignments', Language@internet, *8*: 1–24.

Gershon, I. (2010) *The breakup 2.0: Disconnecting over new media*, Ithaca, NY: Cornell University Press.

Giles, J. (2012) 'Learn a language, translate the web', *New Scientist*, *213*(2847): 18–19.

Goffman, E. (1959) *The presentation of self in everyday life*, New York: Doubleday.

Gorter, D. (2008) 'European minority languages: Endangered or revived', *Endangered languages and language learning*, *12*, 169–175.

Graddol, D. (1997) 'The future of English? A guide to forecasting the popularity of the English language in the 21st century', *English World-Wide*, *20*(1): 171–172.

Graham, M. (2012) 'Big data and the end of theory', *The Guardian*, 9, retrieved from www.theguardian.com/news/datablog/2012/mar/09/big-data-theory (accessed 21 January 2016).

Guilford, G. (2014) Here's why the name of Hong Kong's "Umbrella Movement" is so subversive, retrieved from http://qz.com/283395/how-hong-kongs-umbrella-movement-protesters-are-using-their-native-language-to-push-back-against-beijing/ (accessed 12 August 2015).

Gumperz, J. J. (1982) *Discourse strategies*, Cambridge, UK: Cambridge University Press.

Gupta, A. F. (1997) 'The Internet & the English language', paper presented at the First Conference on Postcolonial Theory, retrieved from https://courses.nus.edu.sg/course/ellibst/poco/paper6.html (accessed 2 December 2015).

Hafner, C., Li, D. and Miller, L. (2015) 'Language choice among peers in project-based learning: A Hong Kong case study of English language learners' Plurilingual practices in out-of-class computer-mediated communication', *The Canadian Modern Language Review*, *71*(4): 441–470.

Halim, N. S. and Maros, M. (2014) 'The functions of code-switching in facebook interactions', *Procedia-Social and Behavioral Sciences*, *118*: 126–133.

Hecht, B. and Gergle, D. (2010) 'The tower of Babel meets web 2.0: User-generated content and its applications in a multilingual context', *Proceedings of the SIGCHI Conference on Human Factors in Computing Systems*, Atlanta, 10–15 April 2010.

Heller, M. (2010) 'The commodification of language', *Annual Review of Anthropology*, *39*: 101–114.

Heller, M. and Duchêne, A. (2012). 'Pride and profit: Changing discourses of language, capital and nation-state', in A. Duchêne and M. Heller (eds) *Language in late capitalism: Pride and profit*, London: Routledge.

The Help Forum on Flickr. (2007) Retrieved from www.flickr.com/help/forum/zh-hk/ (accessed 10 June 2007).

Hendus, U. (2015) '"See translation": Explicit and implicit language policies on Facebook', *Language Policy*, 14(4): 397–417.

Henwood, F., Kennedy, H. and Miller, N. (eds) (2001) *Cyborg lives? Women's technobiographies*, York: Raw Nerve Books.

Herring, S. C. (1997) 'Ethics in cyberresearch: To cite, or not to cite?' *The College*, 1(2): 18–23.

Herring, S. C. (2002) 'Computer-mediated communication on the Internet', *Annual Review of Information Science and Technology*, 36: 109–68.

Herring, S. C. (2004) 'Slouching toward the ordinary: Current trends in computer-mediated communication', *New Media and Society*, 6(1): 26–36.

Herring, S. C. (2009) 'Web content analysis: Expanding the paradigm', in J. Hunsinger, L. Klastrup and M. Allen (eds) *International handbook of Internet research*, London and New York: Springer.

Herring, S. C. (2011) 'Computer-mediated conversation, Part II: Introduction and overview', *Language@Internet*, 8, retrieved from www.languageatinternet.org/articles/2011/Herring (accessed 23 January 2016).

Herring, S. C. (2015) 'New frontiers in interactive multimodal communication', paper presented at the first conference on Approaches to Digital Discourse Analysis, Valencia, Spain, 19–20 November 2015.

Herring, S. C., Paolillo, J. C., Ramos-Vielba, I., Kouper, I., Wright, E., Stoerger, S., Scheidt, L. A. and Clark, B. (2007) 'Language networks on LiveJournal', paper presented at the 40th Annual Hawaii International Conference on System Sciences, Waikoloa, Big Island, Hawaii, 3–6 January 2007.

Higgins, E. T. (1987) 'Self-discrepancy: A theory relating self and affect', *Psychological review*, 94(3): 319–340.

Hinrichs, L. (2006) *Codeswitching on the web: English and Jamaican Creole in e-mail communication*, Amsterdam and Philadelphia: John Benjamins.

Hogan-Brun, G. (2011) 'Language planning and media in minority language and plurilingual contexts', *Current Issues in Language Planning*, 12(3): 325–329.

Honeycutt, C. and Cunliffe, D. (2010) 'The use of the Welsh language on Facebook: An initial investigation', *Information, Communication & Society*, 13(2): 226–248.

Honeycutt, C. and Herring, S.C. (2009) 'Beyond microblogging: Conversation and collaboration via Twitter', paper presented at the 42nd Hawaii International Conference on System Sciences, Waikoloa, Big Island, Hawaii, January 2009.

Hong, L., Convertino, G. and Chi, E. H. (2011) 'Language matters in Twitter: A large scale study', *Proceedings of the Fifth International AAAI Conference on Weblogs and Social Media*, Barcelona, Spain, 17–21 July 2011, retrieved from www.aaai.org/ocs/index.php/ICWSM/ICWSM11/paper/view/2856/3250 (accessed 22 January 2016).

Horner, K. and Krummes, C. (2011) 'Small languages in globalized spaces: Conflicting representations of Luxembourgish on YouTube', paper presented at the Language in the Media Conference, Limerick, 6–8 June 2011.

Huang, D. (2009) 'Language use in asynchronous computer-mediated communication in Taiwan', *Australian Review of Applied Linguistics*, 32(2): 12.1–12.22.

Internet World Stats. (2015) Top ten languages in the Internet, retrieved from www.internetworldstats.com/stats7.htm (accessed 11 June 2015).

Ito, M., Baumer, S., Bittanti, M., Boyd, D., Cody, R. and Herr, B. (2010) *Hanging out, messing around, geeking out: Living and learning with new media*, Cambridge, MA: MIT Press.

Ivković, D. and Lotherington, H. (2009) 'Multilingualism in cyberspace: Conceptualising the virtual linguistic landscape', *International Journal of Multilingualism*, 6(1): 17–36.

Ivanić, R. and Tseng, M. I. L. (2005) *Understanding the relationships between learning and teaching: An analysis of the contribution of applied linguistics*, London: National Research and Development Centre.

Jaffe, A. (2007) 'Minority language movements', in M. Heller (ed) *Bilingualism: A social approach*, New York, UK: Palgrave Macmillan.

James, G. (2001) 'Cantonese particles in Hong Kong students' English e-mails', *English Today*, 17(03): 9–16.

Jaworska, S. (2014) 'Playful language alternation in an online discussion forum: The example of digital code plays', *Journal of Pragmatics*, 71: 56–68.

Jaworski, A. (2015) 'Globalese: A new visual-linguistic register', *Social Semiotics*, 25(2): 217–235.

Johnstone, B. and Baumgardt, D. (2004) '"Pittsburghese" online: Vernacular norming in conversation', *American Speech*, 79(2): 115–145.

Jones, E. H. G. and Uribe-Jongbloed, E. (eds) (2012) *Social media and minority languages: Convergence and the creative industries*, Bristol, UK: Multilingual Matters.

Jones, G. M., Schieffelin, B. B. and Smith, R. E. (2011) 'When friends who talk together stalk together: Online gossip as metacommunication', in C. Thurlow and K. Mroczek (eds) *Digital discourse: Language in the new media*, New York and London: Oxford University Press.

Jørgensen, J. N. (2008) 'Polylingual languaging around and among children and adolescents', *International Journal of Multilingualism*, 5(3): 161–176.

Jørgensen, J. N., Karrebæk, M. S., Madsen, L. M. and Møller, J. S. (2011) 'Polylanguaging in superdiversity', *Diversities*, 13(2): 23–37.

Juffermans, K., Blommaert, J., Kroon, S. and Li, J. (2014) 'Dutch–Chinese repertoires and language ausbau in superdiversity: A view from digital media', *Discourse, Context & Media*, 4: 48–61.

Jurgens, D., Dimitrov, S. and Ruths, D. (2014) 'Twitter users# codeswitch hashtags!# moltoimportante# wow', *Proceedings of the First Workshop on Computational Approaches to Code Switching*, Doha, Qatar, 25–29 October, retrieved from www.aclweb.org/anthology/W14-3906 (accessed 23 January 2016).

Kachru, B. B. (1992) *The other tongue: English across cultures*, Urbana and Chicago: University of Illinois Press.

Karrebæk, M. S., Stæhr, A. and Varis, P. (2015) 'Punjabi at heart: Language, legitimacy, and authenticity on social media', *Discourse, Context & Media, 8*: 20–29.

Katsuno, H. and Yano, C. (2007) 'Kaomoji and expressivity in a Japanese housewives' chatroom', in B. Danet and S. C. Herring (eds) *The multilingual internet: Language, culture, and communication online*, New York and Oxford: Cambridge University Press.

Kelly, C. (2015) *Do you know what I mean: A linguistic study of the understanding of emoticons and emojis in text messages*. Unpublished Bachelor thesis. Stockholm: Halmstad University.

Kelly-Holmes, H. (2004) 'An analysis of the language repertoires of students in higher education and their language choices on the Internet (Ukraine, Poland, Macedonia, Italy, France, Tanzania, Oman and Indonesia)', *International Journal of Multicultural Societies*, 6(1): 29–52.

Kelly-Holmes, H. (2005) *Advertising as multilingual communication*, Basingstoke: Palgrave Macmillan.

Kelly-Holmes, H. (2011) 'Sex, lies and thematising Irish New media, old discourses?', *Journal of Language And Politics, 10*: 511–534.

Kelly-Holmes, H. (2013a) '"Choose your language!" Categorisation and control in cyberspace', *International Yearbook of European Sociolinguistics*, 27(1): 132–145.

Kelly-Holmes, H. (2013b) 'Translation in progress', in S. Pietikainen and H. Kelly-Holmes (eds) *Multilingualism and the periphery*, Oxford, England: Oxford University Press.

Kelly-Holmes, H. (2014) 'Linguistic fetish: The sociolinguistics of visual multilingualism', in D. Machin (ed) *Visual communication*, Berlin: De Gruyter Mouton.

Kelly-Holmes, H. and Milani, T. (eds) (2013) *Thematising multilingualism in the media*, Amsterdam and Philadelphia: John Benjamins.

Kennedy, H. (2003) 'Technobiography: Researching lives, online and off', *Biography*, 26(1): 120–139.

Knobel, M. and Lankshear, C. (2008) 'Digital literacy and participation in online social networking spaces', in C. Lankshear and M. Knobel (eds) *Digital literacies: Concepts, policies and practices*, New York: Peter Lang Publishing.

Koutsogiannis, D. and Mitsikopoulou, B. (2003) 'Greeklish and Greekness: Trends and discourses of "glocalness"', *Journal of Computer-Mediated Communication*, 9(1), retrieved from http://onlinelibrary.wiley.com/doi/10.1111/j.1083-6101.2003.tb00358.x/full (accessed 23 January 2016).

Koutsogiannis, D. and Mitsikopoulou, B. (2007) 'Greeklish and greekness: Trends and discourses of "glocalness"', in B. Danet and S. C. Herring (eds) *The multilingual internet: Language, culture, and communication online*, New York and Oxford: Cambridge University Press.

Kozinets, R. V. (2010) *Netnography*, London: Sage.

Kramsch, C. (2015) 'Afterword: Challenging multilingualism', in U. Jessner-Schmid and C. J. Kramsch (eds) *The multilingual challenge: Cross-disciplinary perspectives*, Berlin: De Gruyter Mouton.

Kress, G. (2009) *Multimodality: A social semiotic approach to contemporary communication*, London and New York: Routledge.

Kress, G. R. and Van Leeuwen, T. (1996) *Reading images: The grammar of visual design*, London and New York: Routledge.

Kytölä, S. (2013) *Multilingual language use and metapragmatic reflexivity in Finnish Internet football forums*, Jyväskylä: University of Jyväskylä.

Labov, W. (1984) 'Field methods of the project on linguistic change and variation', in J. Baugh and J. Sherzer (eds) *Language in use: Readings in sociolinguistics*, Englewood Cliffs, NJ: Prentice Hall.

Lakarnchua, O. (2015) 'Examining the potential of fansubbing as a language learning activity', *Innovation in Language Learning and Teaching*, retrieved from www.tandfonline.com/doi/abs/10.1080/17501229.2015.1016030 (accessed 23 January 2016).

Lam, W. S. E. (2000) 'L2 literacy and the design of the self: A case study of a teenager writing on the Internet', *TESOL Quarterly*, *34*(3): 457–82.

Lam, W. S. E. (2004) 'Second language socialization in a bilingual chat room: Global and local considerations', *Language Learning & Technology*, *8*(3): 44–65.

Lam, W. S. E. (2009) 'Multiliteracies on instant messaging in negotiating local, translocal, and transnational affiliations: A case of an adolescent immigrant', *Reading Research Quarterly*, *44*(4): 377–397.

Lam, W. S. E. (2013) 'Multilingual practices in transnational digital contexts', *TESOL Quarterly*, *47*(4): 820–825.

Lange, P. (2007) 'Publicly private and privately public: Social networking on YouTube', *Journal of Computer–Mediated Communication*, *13*(1), retrieved from www.cs.uwaterloo.ca/~apidduck/CS432/Assignments/YouTube.pdf (accessed 1 May 2012).

Lankshear, C. and Knobel, M. (2008) (eds) *Digital literacies: Concepts, policies and practices.* New York: Peter Lang Publishing.

Lankshear, C. and Knobel, M. (2009) 'More than words: Chris Searle's approach to critical literacy as cultural action', *Race & Class*, *51*(2): 59–78.

Lankshear, C. and Knobel, M. (2011) *New literacies: Changing knowledge and classroom learning*, 3rd edn, Buckingham: Open University Press.

La Rue, F. (2011) Report of the Special Rapporteur on the promotion and protection of the right to freedom of opinion and expression. Human Rights Council, United Nations, retrieved from www2.ohchr.org/english/bodies/hrcouncil/docs/17session/A.HRC.17.27_en.pdf (accessed 2 July 2015).

Lee, C. (2002) 'Literacy practices of computer-mediated communication in Hong Kong', *Reading Matrix*, 2(2) Special Issue: Literacy and the Web, retrieved from www.readingmatrix.com/articles/lee/article.pdf (accessed 1 May 2012).

Lee, C. (2007a) 'Text-making practices beyond the classroom context: Private instant messaging in Hong Kong', *Computers and Composition*, *24*(3): 285–301.

Lee, C. (2007b) 'Linguistic features of email and ICQ instant messaging in Hong Kong', in B. Danet and S. C. Herring (eds) *The multilingual Internet: Language, culture, and communication online*, New York and Oxford: Oxford University Press.

Lee, C. (2007c) 'Affordances and text-making practices in online instant messaging', *Written Communication*, *24*(3): 223–249.

Lee, C. (2011) 'Texts and practices of micro-blogging: Status updates on Facebook', in C. Thurlow and K. Mroczek (eds) *Digital discourse: Language in the new media*, New York and London: Oxford University Press.

Lee, C. (2012) 'Researching the texts and practices in an online photo-sharing site', in M. Sebba, S. Mahootian, and C. Jonsson (eds) *Language mixing and code-switching in writing: Approaches to mixed-language written discourse*, London: Routledge.

Lee, C. (2014) 'Language choice and self-presentation in social media: The case of university students in Hong Kong', in P. Seargeant and C. Tagg (eds) *The language of social media: Community and identity on the internet*, Basingstoke: Palgrave Macmillan.

Lee, C. (2015) 'Digital discourse@Public space: Flows of language online and offline', in R. Jones, A. Chik and C. Hafner (eds) *Discourse and digital practices: Doing discourse analysis in the digital age*, London: Routledge.

Lee, C. and Barton, D. (2011) 'Constructing glocal identities through multilingual writing practices on Flickr.com', *International Multilingual Research Journal*, 5: 1–21.

Lee, C. and Barton, D. (2012) 'Researching the texts and practices in an online photo-sharing site', in M. Sebba, S. Mahootian and C. Jonsson (eds) *Language mixing and code-switching in writing: Approaches to mixed-Language written discourse*, Abingdon: Routledge.

Lee, C. and Chau, D. (2015) 'Tagging the #UmbrellaMovement: Indexing Hongkonger identities through bilingual hashtags on Instagram', paper presented at the 6th International "Language in the media" Conference, University of Hamburg, 7–9 September 2015.

Lee, C. and Lien, P. (2011) 'Facebook group as a third space for teaching and learning', paper presented at the Second International Conference Popular Culture and Education, Hong Kong Institute of Education, 7–10 December 2011.

Lee, H. K. (2011) 'Participatory media fandom: A case study of anime fansubbing', *Media Culture & Society*, 33(8): 1131–1147.

Lee, J. S. (2006), 'Exploring the relationship between electronic literacy and heritage language maintenance', *Language Learning & Technology*, 10(2): 93–113.

Lenihan, A. (2011) '"Join our community of translators": Language ideologies and Facebook', in C. Thurlow and K. Mroczek (eds) *Digital discourse: Language in the new media*, New York and London: Oxford University Press.

Lenihan, A. (2012). 'The interaction of language policy, minority languages & new media: Facebook', paper presented at the Sociolinguistics Symposium 19, Berlin, Germany: Freie Universität Berlin, 21–24 August 2012.

Lenihan, A. (2013) *The Interaction of Language Policy, Minority Languages & New Media: A Study of the Facebook Translations Application*. PhD Thesis, University of Limerick, retrieved from https://ulir.ul.ie/bitstream/handle/10344/3590/Lenihan_2013_interaction.pdf?sequence=5 (assessed 24 January 2016).

Leppänen, S. (2007) 'Youth language in media contexts: Insights into the functions of English in Finland', *World Englishes*, 26(2): 149–169.

Leppänen, S. and Häkkinen, A. (2012) 'Buffalaxed superdiversity: Representations of the Other on YouTube', *Diversities*, 14(2): 17–33.

Leppänen, S. and Peuronen, S. (2012) 'Multilingualism and the Internet', in M. Martin-Jones, A. Blackledge and A. Creese (eds) *The encyclopedia of applied linguistics*, London and New York: Routledge.

Leppänen, S., Pitkänen-Huhta, A., Nikula, T., Kytölä, S., Törmäkangas, T., Nissinen, K., Kääntä, L., Räisänen, T., Laitinen, M., Pahta, P., Koskela, H., Lähdesmäki, S. and Jousmäki, H. (2011) *National survey on the English language in Finland: Uses, meanings and attitudes*, Helsinki: Research Unit for Variation, Contacts and Change in English.

Leppänen, S., Pitkänen-Huhta, A., Piirainen-Marsh, A., Nikula, T. and Peuronen, A. (2009) 'Young people's translocal new media uses: A multiperspective analysis of language choice and heteroglossia', *Journal of Computer-Mediated Communication*, 14(4): 1080–1107.

Lexander, K. V. (2012) 'Analyzing multilingual texting in Senegal: An approach for the study of mixed language SMS', in M. Sebba, S. Mahootian and C. Jonsson (eds) *Language mixing and code-switching in writing*, London: Routledge.

Li, David C. S. (1998) 'Incorporating L1 pragmatic norms and cultural values in L2: Developing English language curriculum for EIL in the Asia-Pacific region', *Asian Englishes*, 1(1): 31–50.

Li, David C. S. (1999) 'Linguistic convergence: Impact of English on Hong Kong Cantonese', *Asian Englishes*, 2(1): 5–36.

Li, David C. S. (2000) 'Cantonese-English code-switching research in Hong Kong: a Y2K review', *World Englishes*, 19(3): 305–322.

Liao, Han-teng. (2015a) Internet users by language, retrieved from http://people.oii.ox.ac.uk/hanteng/2015/02/24/internet-users-by-language-the-top-20/ (accessed 30 July 2015).

Liao, Han-teng. (2015b). 'Language- and Locale-based digital literacy indicators: World's internet for development', Paper presented at the Conference on Digital Humanities, Open University of Hong Kong, December 2015.

Lindholm, L. (2013) 'The maxims of online nicknames', in C. Susan, D. S. Herring and V. Tuija (eds) *Pragmatics of computer-mediated communication*, Berlin: De Gruyter Mouton.

Lippi-Green, R. (1997) *English with an accent: Language, ideology, and discrimination in the United States*, London and New York: Routledge.

Lipski, J. (2005) 'Code-switching or borrowing? No sé so no puedo decir, you know', in L. Sayahi and M. Westmoreland (eds) *Selected proceedings of the second workshop on Spanish sociolinguistics*, Somerville, MA: Cascadilla Proceedings Project.

Liu, J. (2011) 'Deviant writing and youth identity: Representation of dialects with Chinese characters on the internet', *Chinese Language and Discourse: An International and Interdisciplinary Journal*, 2(1): 58–79.

Liu, J. (2012) 'The use of Chinese dialects on the internet: Youth language and local youth identity in urban China', in J. Liu and H. Tao (eds) *Chinese under globalization: Emerging trends in language use in China*, Singapore: World Scientific Publishing.

Luk, J. C. (1998) 'Hong Kong students' awareness of and reactions to accent differences', *Multilingua*, 17(1): 93–106.

Mafu, S. (2004) 'From the oral tradition to the information era: The case of Tanzania', *International Journal on Multicultural Societies*, 6(1): 53–78.

Mair, V. (2011) 'Kung-fu (Gongfu) Tea', *Language Log*, retrieved from http://languagelog.ldc.upenn.edu/nll/?p=3366 (assessed 24 January 2016).

Majumder, P., Mitra, M. and Datta, K. (2006) 'Multilingual information access: An Indian language perspective', *Proceedings of the ACM SIGIR Workshop on New Directions in Multilingual Information Access*, Seattle, 10 August, 2006.

Markman, K. M. and Oshima, S. (2007) 'Pragmatic play? Some possible functions of English emoticons and Japanese kaomoji in computer-mediated discourse', paper presented at the Association of Internet Researchers Annual Conference 8.0: Lets Play, Vancouver, Canada, 17–20 October 2007.

Martin-Jones, M. and Jones, K. E. (eds) (2001) *Multilingual literacies: Reading and writing different worlds*, Philadelphia: John Benjamins Publishing.

McClure, E. (2001) 'Oral and written Assyrian codeswitching', in R. Jacobson (ed) *Codeswitching worldwide II*, Berlin: Mouton de Gruyter.

Merchant, G. (2006) 'Identity, social networks and online communication', *E-Learning and Digital Media*, 3(2): 235–244.

Morris, D., Cunliffe, D. and Prys, C. (2012) 'Social networks and minority languages speakers: the use of social networking sites among young people', *Sociolinguistic Studies*, 6(1): 1–20.

Muysken, P. (1997) 'Code-switching processes: Alternation, insertion, congruent lexicalization', in M. Pütz (ed) *Language choices: Conditions, constraints, and consequences*, Philadelphia: John Benjamins Publishing.

Myers-Scotton, C. (1992) 'Comparing codeswitching and borrowing', *Journal of Multilingual & Multicultural Development*, 13(1–2): 19–39.

Myers-Scotton, C. (1993) *Duelling languages: Grammatical structure in codeswitching.* New York: Oxford University Press.

Myers-Scotton, C. (1997) *Duelling languages: Grammatical structure in codeswitching*, Oxford, UK: Oxford University Press.

Myers-Scotton, C. (2010) 'Patterns and predictions for code-switching with Arabic', in R. Bassiouney (ed) *Arabic and the media: Linguistic analyses and applications*, Netherlands: Hotei Publishing.

Negrón Goldbarg, R. (2009) 'Spanish-English codeswitching in email communication', *Language@Internet*, 6, article 3, retrieved from www.languageatinternet.org/articles/2009/2139 (accessed 26 January 2016).

Nishimura, Y. (2003) 'Linguistic innovations and interactional features of casual online communication in Japanese', *Journal of Computer-Mediated Communication*, 9(1), retrieved from http://onlinelibrary.wiley.com/doi/10.1111/j.1083–6101.2003.tb00356.x/abstract (accessed 26 January 2016).

Nishimura, Y. (2007) 'Linguistic innovations and interactional features in Japanese BBS communication', in B. Danet and S. C. Herring (eds) *The multilingual Internet: Language, culture, and communication online*, Oxford, UK: Oxford University Press.

O'Hagan, M. (2009) 'Evolution of user-generated translation: Fansubs, translation hacking and crowdsourcing', *Journal of Internationalisation and Localisation*, 1(1): 94–121.

O'Hagan, M. (2011) 'Community translation: Translation as a social activity and its possible consequences in the advent of Web 2.0 and beyond', *Linguistica Antverpiensia, New Series–Themes in Translation Studies*, 10, retrieved from https://lans-tts.uantwerpen.be/index.php/LANS-TTS/article/view/275 (assessed 16 January 2016).

O'Reilly, T. (2005) Web 2.0: Compact definition, retrieved from http://radar.oreilly.com/2005/10/web-20-compact-definition.html (assessed 11 January 2016).

Och, F. (2005) 'The machines do the translating', *Google Blog*, retrieved from http://googleblog.blogspot.hk/2005/08/machines-do-translating.html (accessed 6 September 2012).

Olohan, M. (2014) 'Why do you translate? Motivation to volunteer and TED translation', *Translation Studies*, 7(1): 17–33.

Ostler, R. (1999) 'Disappearing languages', *The Futurist*, 33(7): 16–22.

Otsuji, E. and Pennycook, A. (2010) 'Metrolingualism: Fixity, fluidity and language in flux', *International Journal of Multilingualism*, 7(3): 240–254.

Page, R. (2011) *Stories and social media*, London: Routledge.

Page, R. (2012) 'The linguistics of self-branding and micro-celebrity in Twitter: The role of hashtags', *Discourse & Communication*, 6(2): 181–201.

Page, R., Barton, D., Unger, J. W. and Zappavigna, M. (2014) *Researching language and social media: A student guide*, London and New York: Routledge.

Palfreyman, D. and Al-Khalil, M. (2003) '"A funky language for Teenzz to use:" Representing Gulf Arabic in instant messaging', *Journal of Computer-Mediated Communication*, 9(1), retrieved from http://onlinelibrary.wiley.com/doi/10.1111/j.1083–6101.2003.tb00355.x/abstract (assessed 11 January 2016).

Pann, Y. M. and Phatak M. (2012) 'Search engines and Asian languages', in L. Vannini and H. Le Crosnier (eds) *Net. Lang: Towards the multilingual cyberspace*, Caen: C&F Editions. Retrieved from www.unesco.org/new/fileadmin/MULTIMEDIA/HQ/CI/CI/pdf/netlang_EN_pdfedition.pdf (assessed 11 January 2016).

Paolillo, J. C. (2007) 'How much multilingualism? Language diversity on the Internet', in B. Danet and S. C. Herring (eds) *The multilingual internet: Language, culture, and communication online*, New York and Oxford: Oxford University Press.

Papen, U. (2005) *Adult literacy as social practice: More than skills*, London and New York: Routledge.

Pavlenko, A. (2007) 'Autobiographic narratives as data in applied linguistics', *Applied Linguistics*, 28(2): 163–88.

Pavlenko, A. and Blackledge, A. (eds) (2004) *Negotiation of identities in multilingual contexts*, Clevedon, UK: Multilingual Matters.

Pérez-Ortiz, J. A. (2011) 'Social translation: How massive online collaboration could take machine translation to the next level', *Proceedings of the Second European Language Resources and Technologies Forum: Language Resources of the Future – The Future of Language Resources*, Barcelona, Spain, 11–12 February 2010.

Peuronen, S. (2011) '"Ride hard, live forever": Translocal identities in an online community of extreme sports Christians', in C. Thurlow and K. Mroczek (eds) *Digital discourse: Language in the new media*, Oxford: Oxford University Press.

Phillipson, R. (1992) *Linguistic imperialism*, Oxford: Oxford University Press.

Pollara, P. and Zhu, J. (2011) 'Social networking and education: Using Facebook as an edusocial space', *Proceedings of the Society for Information Technology & Teacher Education International Conference*, retrieved from www.editlib.org/d/36833/ (assessed 27 January 2016).

Prado, D. (2012) 'Language presence in the real world and cyberspace', in L. Vannini and H. Le Crosnier (eds) *Net. Lang: Towards the multilingual cyberspace*, Caen: C&F Editions. Retrieved from www.unesco.org/new/fileadmin/MULTIMEDIA/HQ/CI/CI/pdf/netlang_EN_pdfedition.pdf (assessed 11 January 2016).

Preston, D. R. (2004) 'Three kinds of sociolinguistics', in C. Fought (ed) *Sociolinguistic variation: Critical reflections*, Oxford: Oxford University Press.

Provine, R. R., Spencer, R. J. and Mandell, D. L. (2007) 'Emotional expression online emoticons punctuate website text messages', *Journal of Language and Social Psychology*, 26(3), 299–307.

Raley, R. (2003) 'Machine translation and global English', *The Yale Journal of Criticism*, *16*(2): 291–313.

Rau, D. V. and Yang, M. C. (2009) 'Digital transmission of language and culture', in M. Florey (ed) *Language endangerment and maintenance in the austronesian region*, Oxford, UK: Oxford University Press.

Reid, J. (2011) '"We don't Twitter, we Facebook": An alternative pedagogical space that enables critical practices in relation to writing', *English Teaching: Practice and Critique*, *10*(1): 58–80.

SCMP. (2011) 'Mormon wins fans with humorous take on Cantonese slang', *South China Morning Post*, retrieved from www.scmp.com/article/964716/mormon-wins-fans-humorous-take-cantonese-slang (accessed 20 August 2014).

SCMP. (2014). Is Cantonese in danger? Hongkongers take steps to protect their heritage, retrieved from www.scmp.com/lifestyle/family-education/article/1450856/hongkongers-take-steps-preserve-their-language-and (accessed 14 June 2016).

SCMP. (2016). 'Hongkongers mix English and Cantonese into new language, Kongish', *South China Morning Post*, retrieved from www.scmp.com/lifestyle/article/1903452/hongkongers-mix-english-and-cantonese-new-language-kongish (accessed 20 June 2016).

Scollon, R. and Scollon, S. W. (2003) *Discourses in place: Language in the material world*, London and New York: Routledge.

Seargeant, P. and Tagg, C. (2011) 'English on the internet and a "post-varieties" approach to language', *World Englishes*, *30*(4): 496–514.

Seargeant, P., Tagg, C. and Ngampramuan, W. (2012) 'Language choice and addressivity strategies in Thai-English social network interactions', *Journal of Sociolinguistics*, *16*(4): 510–531.

Sebba, M. (2012) 'Researching and theorising multilingual texts', in M. Sebba, S. Mahootian and C. Jonsson (eds) *Language mixing and code-switching in writing*, London: Routledge.

Sebba, M. (2015) 'Iconisation, attribution and branding in orthography', *Written Language & Literacy*, *18*(2): 208–227.

Siebenhaar, B. (2006) 'Code choice and code-switching in Swiss-German Internet Relay Chat rooms', *Journal of Sociolinguistics*, *10*(4): 481–506.

Sindoni, M. G. (2013) *Spoken and written discourse in online interactions: A multimodal approach*, London and New York: Routledge.

Sindoni, M. G. (2014) 'Through the looking glass: A social semiotic and linguistic perspective on the study of video chats', *Text & Talk*, *34*(3): 325–347.

Small, T. A. (2011) 'What the hashtag? A content analysis of Canadian politics on Twitter', *Information, Communication & Society*, *14*(6): 872–895.

Snow, D. (2004) *Cantonese as written language: The growth of a written Chinese vernacular*, Hong Kong: Hong Kong University Press.

Snyder, I. A. (2015) 'Discourses of "curation" in digital times', in R. H. Jones, A. Chik and C. A. Hafner (eds) *Discourse and digital practices: Doing discourse analysis in the digital age*, London and New York: Routledge.

Specter, M. (1996) 'World, wide, web: 3 English words', *The New York Times*, 14 April, Section 4, 1.

Sperlich, W. (2005) 'Will cyberforums save endangered languages? A Niuean case study', *International Journal of the Sociology of Language*, *172*: 51–77.

Spilioti, T. (2009) 'Graphemic representation of text-messaging: Alphabet-choice and code-switches in Greek SMS', *Pragmatics*, *19*(3): 393–412.

Su, H-Y. (2007) 'The multilingual multiorthographic Taiwan-based Internet', in B. Danet and S. C Herring (eds) *The multilingual Internet: Language, culture and communication online*, New York: Oxford University Press.

Tagg, C. and Seargeant, P. (2012) 'Writing systems at play in Thai-English online interactions', *Writing Systems Research*, *4*(2): 195–213.

Temple, B. (2006) 'Being bilingual: Issues for cross-language research', *Journal of Research Practice*, *2*(1): 1–15.

Themistocleous, C. (2012) 'Cypriot Greek Nuclear Pitch Accents', in Z. Gavriilidou, A. Efthymiou, E. Thomadaki and P. Kambakis-Vougiouklis (eds) *Selected papers of the 10th ICGL*, Komotini/Greece: Democritus University of Thrace.

Themistocleous, C. (2015) 'Digital code-switching between Cypriot and Standard Greek: Performance and identity play online', *International Journal of Bilingualism*, 19(3): 282–297.

Thomas, G. (1991) *Linguistic purism*, London and New York: Longman Publishing Group.

Thomsen, E. B. (2002) 'Preserving linguistic diversity: Endangered languages on the world wide web', *Collection Building*, 21(3): 134–135.

Thorne, S. L. and Ivković, D. (2015) 'Multilingual Eurovision meets plurilingual YouTube', in D. A. Koike and C. S. Blyth (eds) *Dialogue in multilingual and multimodal communities*, Amsterdam and Philadelphia: John Benjamins.

Thurlow, C. (2012) 'Determined creativity: Language play in new media discourse', in R. Jones (ed) *Discourse and creativity*, London: Pearson.

Tiidenberg, K. and Cruz, E. G. (2015) 'Selfies, Image and the Re-making of the Body', *Body & Society*, 21(4): 77–102.

Times Higher Education. (2015) Why I asked Twitter about academics from less privileged backgrounds, retrieved from www.timeshighereducation.com/blog/why-i-asked-twitter-about-academics-less-privileged-backgrounds (accessed 1 March 2016).

Tseliga, T. (2007) '"It's all Greeklish to Me!" Linguistic and sociocultural perspectives on Greeklish (Roman-alphabeted Greek) in asynchronous CMC', in B. Danet and S. C Herring (eds) *The multilingual Internet: Language, culture and communication online*, New York: Oxford University Press.

Vaisman, C. (2011) 'Performing girlhood through typographic play in Hebrew blogs', in C. Thurlow and K. Mroczek (eds) *Digital discourse: Language in the new media*, Oxford, UK: Oxford University Press.

Vandekerckhove, R. and Nobels, J. (2010) 'Code eclecticism: Linguistic variation and code alternation in the chat language of Flemish teenagers', *Journal of Sociolinguistics*, 14(5): 657–677.

Varis, P., Wang, X. and Du, C. (2011) 'Identity repertoires on the Internet: Opportunities and constraints', *Applied Linguistics Review*, 2: 265–284.

Vertovec, S. (2010) 'Towards post-multiculturalism? Changing communities, conditions and contexts of diversity', *International Social Science Journal*, 61: 83–95.

W3techs. (2015) Usage of content languages for websites, retrieved from http://w3techs.com/technologies/overview/content_language/all (accessed 6 September 2015).

Warschauer, M. (1998) 'Technology and indigenous language revitalization: Analyzing the experience of Hawaii', *Canadian Modern Language Review*, 55(1): 139–159.

Warschauer, M., El Said, G. and Zohry, A. (2007) 'Language choice online: Globalization and Identity in Egypt', in B. Danet and S. Herring (eds) *The multilingual internet: Language, culture, and communication online*, Oxford, UK: Oxford University Press.

Webstagram. (2014) Top 100 hashtags, retrieved from https://websta.me/hot (accessed 8 August 2014).

Wellman, B. (2001) 'Physical place and cyberplace: Rise of personalized networking', *International Journal of Urban and Regional Research*, 25(2): 227–52.

Wikimedia. (2015) Language proposal policy, retrieved from http://meta.wikimedia.org/wiki/Language_proposal_policy (accessed 6 September 2012).

Wikipedia. (2015) Emoticon, retrieved from https://en.wikipedia.org/wiki/Emoticon#Orz (accessed 6 September 2015).

Wong, N., Tsang, A. and Lok, P. (2016) 'Positioning Kongish in Hong Kong English: A case study of Kongish Daily', paper presented at the 6th Hong Kong Association for Applied Linguistics Conference, 11 June 2016.

The Worldmapper. (2006) retrieved from www.worldmapper.org.

Wright, H. K. (1996) 'Email in African studies', *Convergence: The Journal of Research into New Media Technologies*, 2(1): 19–29.

Yang, C., Lin, K. H. Y. and Chen, H. H. (2007) 'Emotion classification using web blog corpora', in T.Y. Lin, L. Haas, J. Kacprzyk, R. Motwani, A. Broder and H. Ho (eds) *Proceedings of the International Conference on Web Intelligence*, Silicon Valley, California, USA, 2–5 November 2007, retrieved from http://ieeexplore.ieee.org/stamp/stamp.jsp?tp=&arnumber=4427100 (accessed 27 January 2016).

You, X. (2011) 'Chinese white-collar workers and multilingual creativity in the diaspora', *World Englishes*, *30*(3): 409–427.

Yus, F. (2014) 'Not all emoticons are created equal', *Linguagem em (Dis) curso*, *14*(3): 511–529.

Zappavigna, M. (2015) 'Searchable talk: The linguistic functions of hashtags', *Social Semiotics*, *25*(3): 274–291.

Zhang, W. (2012) 'Chinese-English code-mixing among China's netizens', *English Today*, *28*(03): 40–52.

Zhang, W. (2015) 'Multilingual creativity on China's Internet', *World Englishes*, *34*(2): 231–246.

Index

Note: Page numbers in italics indicate figures and tables.